The Quest for Ultra Performance in Liquid Chromatography

Origins of UPLC Technology

Patrick D. McDonald, Ph.D.

with a contribution by Uwe D. Neue, Ph.D.

THE SCIENCE OF WHAT'S POSSIBLE.™

Table of Contents

*Note: Throughout this book, you will find that certain words are highlighted in red at their first occurrence in the text. Definitions and detailed explanations of these terms are found in this illustrated glossary.

List of Figures and Tables

Foreword

Welcome to the first in our series of books on UPLC technology. We want you to know the *why* and the *how* of UPLC principles so that you can apply them with *confidence in* your laboratory to *enhance* your analytical productivity and scientific progress. Whether you are just curious about this new capability, confused about conflicting claims and promises, considering an investment in a new liquid chromatography [LC] system, or are now the owner of a UPLC system, we hope that our books will enable you to lead your colleagues in the successful implementation of ultra performance LC in your workplace.

This volume presents a brief review of the origin and history of LC, showing how early the concepts of ultra performance were recognized and how many decades it took to reduce them to practice. In other volumes, you will learn the relative importance of both chemical separation power [selectivity] and mechanical separation power [efficiency]. You will appreciate why holistic design and innovation were necessary to meet the engineering challenges in minimizing the band-spreading characteristics and volume of a UPLC system operated at very high pressure. You will see how UPLC technology overcomes the limitations of HPLC to achieve new levels of speed without compromising sensitivity and resolution.

Our UPLC books assume that you have some familiarity with chromatography. If you are new to LC, you may wish to read first our companion volume.[1]

We hope that you and your colleagues find great value in our books and refer to them often. We trust they will induce understanding that enables you not only to speak, but to *think* in, the *language* of LC. We welcome your comments, suggestions, and feedback. Thank you and best wishes for success with ultra-performance liquid chromatography!

Joe Arsenault, Eric Grumbach, Doug McCabe, Pat McDonald
Milford, Massachusetts, June 2009

Acknowledgments

Undertaking the creation of a volume such as this within the scope of already crowded work schedules requires teammates with a high level of skill, commitment, support, encouragement, and dedication. In particular, I wish to thank my colleagues: Joseph Arsenault, Dr. Mark Baynham [the champion of this project], Dr. Diane Diehl, Dr. Geoff Gerhardt, Eric Grumbach, Pamela Iraneta, Douglas McCabe, Damian Morrison, Dr. Uwe Neue, and Dr. Thomas Walter for their encouragement, critical reading of the manuscript, and many helpful suggestions, in particular Dr. Neue's substantive contribution to Chapter 10; Dr. Michael Holdoway and Dr. Richard Vivilecchia for sharing recollections of their pioneering work in the creation of HPLC stationary phases in the early 1970s; Dr. Neue and Bernie Monaghan for sharing first-hand experiences from that same period; colleagues Dr. Keith Fadgen, Charles Pidacks, and Douglas Wittmer for sharing both early and recent history; Prof. Vadim Davankov for his generous gift of a copy of his translation of Evgenia Senchenkova's biography of Michael Tswett; Carrie Blake and Erik Goulet, Waters Chemistry R&D Analytical Laboratory, for providing electron and optical photomicrographs of particles; Carla Clayton and Maureen Allegrezza, Waters Information Center, for procuring many hard-to-find references; Roula Ginis and members of her world-class Chemistry Operations marketing communications team, especially: Ekaterini Kakouros and Ian Hanslope for their superb illustrations, design, and layout of the manuscript; and Dawn Maheu for meticulously editing the book and coordinating all aspects of its production.

Dedication In Memoriam

Dr. Marianna Kele

*b. 14 September 1959,
Furta, Hungary*

*d. 1 September 2008,
Budapest, Hungary*

This volume is dedicated to our friend and colleague, Dr. Marianna Kele. It is no accident that she is #1 in the team photo [rear cover] for she forged a strong technical bridge between chromatographic principles and engineering execution that formed a foundation for the creation and enhanced performance of the ACQUITY UPLC® System. Her efforts epitomized interdisciplinary teamwork, and her eloquence in promoting UPLC technology was inspiring.

1979–1983:	Moscow University [Institute of Crude Oil and Natural Gas Manufacturing]
1985:	B.A., Chemistry, M.A. Chemical Engineering, U. Veszprém, Hungary
1985–1987:	Supervisor, Quality Control, Gas Chromatography Method Development Laboratory, Danube Oil Refinery, Százhalombatta, Hungary
1987:	Married Ferenc Adam
1987–1995:	Chromatographer at Medical University, Pécs, Hungary
1989:	Son Gergely born 4 May
1995–2000:	Doctoral Studies [mentors: Robert Ohmacht, U. Veszprém; Georges Guiochon, University of Tennessee, Knoxville]
2000:	Ph.D., Analytical Chemistry, U. Veszprém: Thesis: *Repeatability and Reproducibility of Retention Data and Band Profiles on RPLC Columns* [six publications in *J. Chromatogr. A* based upon this thesis have received 333 citations as of April 2009]
2001–2008:	Senior Research Chemist, Senior Scientist, Manager of Biopharmaceutical Evaluation Group, Waters Corporation, Milford, Massachusetts
2008:	Finalist, LC-GC Emerging Leader in Chromatography Award Co-author of 22 peer-reviewed, highly cited publications Frequent invited speaker at international symposia

1. Revolution—The Invention of Chromatography

An essential condition for all fruitful research is to have at one's disposal a satisfactory technique. All scientific progress is a progress in a method.—Mikhail Tswett, *Thesis*, 1910[2]

Mikhail Tswett began studying the adsorption of plant pigments by various powders at the dawn of the 20th century. A revolution in separation science began in that defining moment when he made a change in his experimental protocol. Unlike the dozens of experiments he had previously performed by mixing a variety of adsorbents with solutions of extracted plant pigments, he chose instead to pack his powder [precipitated calcium carbonate] *"firmly"* into a bed within a glass tube and then pass a solution of extracted pigments through it.[3] As represented in Figure 1-1, this constituted the first LC experiment.

Figure 1–2: Two photographs of Mikhail Semenovich Tswett [b. 14 May 1872, Asti, Italy; d. 26 June 1919, Voronezh, Russia]. Left: taken in Kiel, 1905; right: near his chromatography apparatus, Botany Laboratory, Warsaw Polytechnical Institute, 1913. [Reproduced with permission from reference 6b.]

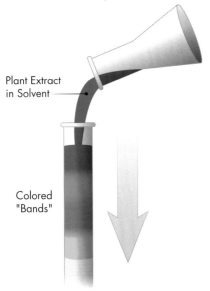

Plant Extract in Solvent

Colored "Bands"

Figure 1–1: The first successful *chromatography* experiment, reported in a 1903 lecture, separated plant pigments. Intense colors characteristic of each compound made it easy for Tswett to detect the progress of the separation and inspired him to combine the Greek words for *color* [*chroma*], *written* [*graphos*] and *writing* [*gramma*] to create terms to describe his new technique. [See Tswett's original sketches in Figure 1–3.]

In 1992, Veronika Meyer summarized the known facts about Tswett's columns and then, using modern LC theory, speculated on the performance of his separations.[4] Tswett's own notes state that his glass columns were 2–3 mm ID and contained a packed bed, only 20–30 mm in length, of *"angular or round"* granules with a mean particle diameter of 0.005 cm [50 μm]. He connected up to five columns in parallel to a single manifold that he could pressurize by squeezing a rubber bulb. In this way, he generated an operating pressure, measured with a manometer, of about one-third of an atmosphere [0.33 bar, not quite 5 psi]. Veronika estimated his mobile-phase flow rate to be about 2–4 mL/min, the time to elute an unretained peak, t_0, only 2.5 sec, and the linear flow velocity, 10 mm/sec—by modern standards, very fast chromatography![5] She determined by reproducing his experiments that his chlorophyll separation required less than 100 theoretical plates. This indicates that the separation factor was ≥ 2 for adjacent bands that were distinctly separated [see Chapter 4].

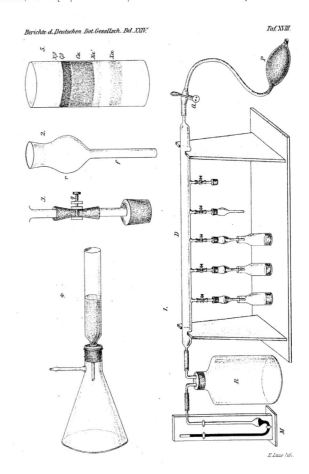

Figure 1–3: Lithograph Plate XVIII drawn by E. Laue for Tswett's 1906 paper[8b] to illustrate his apparatus for chromatography: [1] a manually pressurized manifold for simultaneous operation of up to [2] five small glass columns [2–3 mm i.d.] with integral solvent reservoirs; [3] a pinch clamp was used to control column inlet pressure and thereby flow rate; [4] flow through a larger open glass preparative LC column [10–20 mm i.d.] was assisted by vacuum applied to the side arm of the filter flask; [5] sketch of colored bands for xanthophylls and chlorophylls. Note the uniformity of band shapes, indicative of the skill with which Tswett packed and operated his columns.

We may marvel in hindsight at how this pioneer's great experimental skill surmounted what we now recognize as a severe loss of efficiency because, unknown to Tswett at that time, his sample size had overloaded his column, and his volume flow rate was >100 × faster than the optimum for his particle diameter.

It may seem merely fortuitous that Tswett chose for his first test a mixture that was easy to separate. However, it seems more likely that Tswett, characterized by those who knew him as a highly curious, brilliant, determined, meticulous investigator and lucid lecturer,[6] being thoroughly familiar with 19th century chemical literature in his area of interest, must have instinctively felt that the problem on which he chose to work would ultimately yield important knowledge and have significant impact on the future of science.

2. Evolution—Acceptance of the Chromatographic Technique

Though Tswett died at a young age, he left a legacy in his masterful Warsaw Ph.D. thesis, published as a book in 1910.[2,7] The adoption of his chromatographic technique was slowed for more than two decades since few had access to Tswett's tome or could read it in Russian. One who could was Warsaw native, Wladyslaw Franciszek de Rogowski. He described his successful confirmation of Tswett's chlorophyll separations in his 1912 doctoral thesis under Prof. Charles Dhéré in Fribourg, Switzerland. It certainly did not help that students of the then renowned Herr Prof. Richard Willstätter [1915 Nobel Prize in Chemistry for chlorophyll studies] in Zürich and Berlin did not read Tswett's book carefully and chose an inappropriate packing material for their chromatography of pigments. Not only did they fail to duplicate Tswett's separation of chlorophylls *a* and *b*; their active adsorbent chemically altered the sample components. Willstätter continued to publish his doubts about the preparative utility of LC for 15 years![8]

Independently, Leroy Palmer, having read Tswett's initial reports in a German botany journal,[3] used LC in his doctoral studies on carotenoids from 1910–1913 with Prof. C. Eckles at the University of Missouri. In 1929, Palmer published a monograph on carotenoids in which he described the use of LC and referenced Tswett's papers and thesis. In turn, a young doctoral student, Edgar Lederer, read Palmer's book shortly after he joined Prof. Richard Kuhn's group in Heidelberg, Germany, to study carotenoid chemistry. Seeing the reference to Tswett's thesis, Kuhn remembered that his mentor, Prof. Willstätter, had given him years earlier a private translation of Tswett's book. He managed to find it, and, with its guidance, Lederer went on to perform "*spectacular*" LC separations of plant pigments. This work, published in 1931,[9] was later characterized by Lederer as the renaissance of Tswett's chromatographic method.[10]

Chromatography was put to good use in Kuhn's research group by subsequent students such as Winterstein and Brockmann.[11] It spread among the natural products chemistry community to many laboratories including those of Paul Karrer in Zürich[12] and László Zechmeister in Pécs, Hungary.[13] The study of plant pigments, vitamins, and steroid hormones reached new heights in the 1930s and 1940s. Switzerland was a haven during WWII for researchers such as Croatian Lavoslav [Leopold] Ružička at the Eidgenössische Technische Hochschule [ETH] in Zürich and his assistant Tadeus Reichstein, who, after isolating and identifying hundreds of compounds in coffee[14] and chicory, moved to Basel in 1938. There he and his Australian postdoctoral fellow Charles William Shoppee standardized the practice of open column LC on alumina.[15] LC was the significant tool that enabled these chemists to win Nobel prizes for their work on carotenoids [Kuhn, chemistry,

1938], vitamins [Karrer, chemistry, 1937], terpenes [Ružička, chemistry, 1939] and steroid hormones [Reichstein, medicine, 1950].[16]

3. Revolution—Foundation of Modern LC

Reichstein and his colleagues developed their guidelines for choosing column size, bed dimensions, and weight of adsorbent *via* empirical experiments in which they varied the sample load and determined the general degree of difficulty for many separations. Across the channel, at about the same time, two chemists at the Wool Industries Research Association in Leeds [UK], working on amino-acid analysis, architected a framework for the practice, performance criteria, and theoretical understanding of LC.

Archer J.P. Martin and Richard L.M. Synge received the 1952 Nobel Prize in Chemistry for their development of partition chromatography. In his award lecture, Martin, perhaps paraphrasing Pasteur,[17] said, "*Partition chromatography resulted from the marrying of two techniques, that of chromatography and that of countercurrent solvent extraction. All of the ideas are simple and had people's minds been directed that way the method would have flourished perhaps a century earlier.*"[18]

While still a schoolboy, Martin had developed an interest in fractional distillation, learning about the plates or discrete physical stages used to amplify the efficiency and separation power of distillation columns. At Cambridge, he "*became interested in countercurrent separations and plate theory.*" A year after graduation, he joined the Dunn Nutritional Laboratory at Cambridge, and, pursuing his interest in vitamins, he decided to search for the then unknown Vitamin E. Other colleagues were studying carotenes. In 1933 "*Dr. Winterstein from Richard Kuhn's laboratory in Heidelberg visited us and demonstrated a chromatogram of a crude carotene solution on a chalk column; the carotene separated appropriately into bands of various colours. I was fascinated to see the relationship between the chromatogram and distillation columns and to realize that the processes involved in the separation of the carotenes and of volatile substances were similar; there was relative movement of the two phases and it was their interaction at many points that gave rise to good separations. ... I started separating carotene by distribution between two solvents using separating funnels. I was sufficiently mathematically inclined to work out the extent of separation that can be obtained in this way, and was appalled to find how small this was with a single extraction. So, I set up chains of separating funnels, moving upper and lower layers countercurrently, but found that even when one has such a small number as, say, six funnels, just shaking and separating the layers becomes a full-time job.*"[19] Later, inspired by a single sentence in an American research paper on Vitamin E, Martin built an apparatus for countercurrent fractional extraction.[20] In 1937 he was introduced to Synge who shared his interest in amino acid separations. Synge followed Martin to Leeds where they set out to improve their 45-stage extraction machine.

They were able to determine fatty amino acids in wool, but their machine proved "*troublesome to use and required nursing day and night for several days at a time.*" It was then that Martin "*had an idea of a radically different kind. This was to pack a tube....*" His attempt to create parallel streams of chloroform and water through a bed of wool [hydrophobic] and cotton [hydrophilic] fibers was unsuccessful. Fortunately, Martin persisted in his revolutionary thinking.

Figure 3–1: Young Archer J.P. Martin in his laboratory at Cambridge, looking at some distillation apparatus, ca. 1930. [Photograph used with permission.[21]]

"Up to this time my thinking had been dominated by the idea of moving the two phases in opposite directions simultaneously. But in considering the failure of the cotton and wool experiment I realized that it should be comparatively simple to hold one phase stationary and move only one. The arrangement would then be essentially a chromatogram. I put this idea to Synge and he was full of enthusiasm, and we decided to work on it at once."[18]

In 1941 they published *"A New Form of Chromatogram Employing Two Liquid Phases,"*[22] an elegant exposition of an *"approximate theory of chromatographic separations"*—broadly applicable to all forms of chromatography—and the reduction to practice of a new LC mode. This seminal paper is characterized by a depth of critical thinking and foresight that seems rare in today's publish-or-perish world. As a citation classic, it stands as a model for all contemporary science students. In the realm of LC, we recognize it as a significant scientific milestone for four reasons:

1. In the Introduction, they conceived *gas*-liquid partition chromatography [GC]—*"The mobile phase need not be a liquid but may be a vapour."*[23] Eleven years later, Martin and James first reduced this idea to practice, initiating explosive growth of a technique that became both a scientific and commercial success.[24]

2. In Part I, *A Theory of Chromatography,* by analogy to the physical elements in distillation columns, they introduced the concept of virtual or theoretical plates [and, by extension, the H.E.T.P.—*"height equivalent to one theoretical plate"*] as a measure of column performance in LC [see Efficiency and Plate Number in Appendix].[23] The smaller the H.E.T.P., the more plates there are in a column, and the higher is its efficiency.

3. Also in Part I, they developed a mathematical treatment of band spreading during an analyte's passage through a column, as indicated by an increase in the H.E.T.P., and thereby, in essence, *presaged HPLC and UPLC performance:* *"The height equivalent to a theoretical plate depends upon the factors controlling diffusion and upon the rate of flow of the liquid. There is an optimum rate of flow in any given case, since diffusion from plate to plate becomes relatively more important the slower the flow of liquid and tends of course always to increase the H.E.T.P. Apart from this, the H.E.T.P. is proportional to the rate of flow of liquid and to the square of the particle diameter. Thus the smallest H.E.T.P. should be obtainable by using very small particles and a high pressure difference across the length of the column. The H.E.T.P. depends also on the diffusibility of the*

solute in the solvent employed, and in the case of large molecules, such as proteins, this will result in serious decrease in efficiency as compared with solutes of molecular weights of the order of hundreds."[25]

4. And, perhaps most significantly, they recognized the *practical limitations* of proving their theory: *"The separations obtainable in practice are less than the theory predicts for two principal reasons. First, the partition coefficient is seldom a constant, usually decreasing as the solution becomes stronger. This results in the front of the band becoming steeper, and the back flatter and, more importantly, in the band becoming wider, since the concentrated part moves faster than the dilute part.... The other great source of loss of efficiency lies in lack of uniformity of flow through the column. This lack of uniformity often prevents good separations being realized, even though the solutes be separated in the column itself .... In striving for conditions for uniformity of flow, the high pressure and small particle size desirable for smallest H.E.T.P. have to be abandoned."*[27] They knew what to do to achieve high efficiency, but they did not know how to make and pack uniform small particle columns, nor could they construct suitable instrumentation. For this feat of engineering art and science, the world had to wait for more than two decades.

4. Achieving a Separation— Importance of the Separation Factor [α]

It becomes all too easy in reviewing the history of LC development to form the impression that a low plate height is paramount. Before we get too deep into our discussion of efficiency, we need to realign our perspective to focus on the primary goal of chromatography: to achieve a separation. A corollary to this goal is time optimization. This explains, as Giddings has stated, *"the basis of the change in direction taken in the late 1960s in the development of HPLC. ... the replacement of large particles by small particles ... and low pressure drops by high pressure drops ... has led to the development of an efficient high-speed separation tool of almost universal applicability."*[26] Let us emphasize once more that the quest for ultimate performance in LC remains focused on two main objectives: *achieve the desired separation as quickly as possible.* You will see in subsequent chapters how, with a further reduction in particle size and increase in pressure drop, UPLC technology takes a dramatic step up in LC performance—while the important principles of achieving LC separation and retention remain the same.

Chromatography is a dynamic equilibrium process. At the molecular level, it involves the distribution of solute molecules between a mobile phase and a stationary phase. A tiny difference in the relative affinities of similar, though distinct, solutes for the two phases, when multiplied by the multitude of phase transfers that occur during the passage of molecules through a chromatographic column, accounts for the measurable difference in retention between sample components. Selecting an optimum combination of mobile and stationary phases to maximize the relative differences between their affinities for sample components will have the *most significant effect* on the success of any LC separation.

A detailed description of the thermodynamic and kinetic treatments of the chromatographic process can be found in several reference works[26,27] and are beyond the scope of this essay. Two points, though, must be emphasized:[28]

First, the separation factor, α, is derived from thermodynamic equilibrium theory and ultimately may be represented by the ratio of two retention factors, k_2/k_1, measured under identical conditions, for a given pair of compounds.[29]

The retention factor, k, is simply the ratio of the amount of a compound in the stationary phase to the amount of that compound in the mobile phase at equilibrium:

$$k = \frac{\text{amount of compound in stationary phase}}{\text{amount of compound in mobile phase}}$$

Second, it is quite easy to determine α and k either from static equilibrium measurements or, as shown in Figure 4-1, from chromatographic experiments. The unretained peak represents a marker solute that passes through the chromatographic system without interacting in any significant way with the stationary phase. The volume of mobile phase required to elute this compound from the column bed is called the *void volume*, V_0 [see Column Volume in Appendix]. In practice, what is usually measured, the *hold-up volume*, or V_M, is equal to the volume of mobile phase contained by the chromatographic system from point of injection to point of detection [this includes the collective volume of the injector, column packing (both between and within porous particles—the void volume), column end fittings, connection tubing, and detector cell]. Fortunately, when calculating the separation factor, α, the non-thermodyamically relevant volume contributed by system elements other than the packed bed is canceled out.

Retention volumes V_1 and V_2, are the volumes of mobile phase measured from point of injection to the respective peak maxima for compounds 1 and 2. Each peak in the chromatogram represents directly or indirectly a profile of the concentration of a solute in the mobile phase as it passes the *detection point*, a short but not insignificant distance downstream from the outlet of the chromatographic bed. If the rate of flow of mobile phase [volume per unit time] through the system is constant, then the *volume* axis in a chromatogram can also be represented as a *time* axis by simply dividing each of the volume axis values by the flow rate [see Figure 4–1].

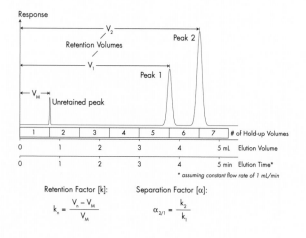

Figure 4–1: How to calculate retention factors [k] and separation factor [α] from measurements on an analytical chromatogram.[30]

Ultimately, the *affinity* of a compound for the stationary phase, relative to its affinity for the mobile phase, determines how much *time* that compound spends in the chromatographic bed. This relative affinity is fixed by the respective chemical nature of the three elements of a particular chromatographic system

[the mobile phase, the stationary phase, and the compound] and the experimental conditions [especially, temperature]. Increasing the volume flow rate of the mobile phase in order to complete a separation in less time does not change the number of hold-up volumes required to elute a particular compound in a particular chromatographic system.

Plate number [N] required for each component at various separation factors [α] to achieve specific resolution values [R_s]

α	N [Calculated using equation in Figure A–7]					
	R_s = 0.6	0.7	0.8	1.0	1.25	1.5
10	2.6	3.5	4.6	7.1	11.1	16.0
5	4	6	8	12	19	28
3	9	12	16	25	39	56
2	23	31	41	64	100	144
1.7	40	55	72	112	175	251
1.5	71	96	125	196	306	441
1.3	174	237	310	484	756	1089
1.2	369	502	655	1024	1600	2304
1.1	1384	1884	2460	3844	6006	8649
1.05	5358	7293	9526	14884	23256	33489
1.01	130,465	177,578	231,938	362,404	566,256	815,409
1.005	520,129	707,954	924,675	1,444,804	2,257,506	3,250,809

Other conditions: k_1 = 2.0; a = 16 [tangent method, see Figure 5–1]

Table 4–1: Baseline separation [$R_s \geq 1.25$] is achieved in a separation of two adjacent components [approximately equal in peak area and detector response, assuming Gaussian peak shape and analytical sample load] with a surprisingly small plate count if the separation factor [α] is high. When α > 3, separation may be achieved by step gradients in short open columns or solid-phase extraction cartridges. The separation capability limit of isocratic separations in classical open columns is at about α ≥ 1.5. Thin-layer chromatography [see Chapter 6] can lower the limit to about α ≥ 1.3. Modern HPLC took separation capability into the realm of α < 1.3.[31] Some researchers have set out to achieve 1,000,000 or more plates so as to effect separations with an α of ≤ 1.01—Why? Because they can![32] [The price paid for this level of efficiency is a separation time measured in hours, not minutes or seconds. See Chapter 10] However, considering the data in this table, a greater return on research investment might well be obtained by improving the selectivity of the separation system—consider the vast difference in efficiency requirements if α can be raised from 1.01 to only 1.05! In this way, the goals of LC cited above—achieving the desired separation as quickly as possible—may be attained.

As can be seen from a chromatogram such as that in Figure 4–1, the larger the separation factor, α, the greater is the *distance between peak centers*. As you will learn in the next section, higher efficiency reduces the *width* of the peaks, increasing their apparent *resolution*, but not their *relative separation*. If α is increased, then a separation may be achieved even with broad peaks [see Table 4–1]. Thus, a higher α value lessens the need for maximum efficiency. To reach the first objective of LC, *achieving a separation*, α is the most powerful tool. To attain the second objective, *highest separation speed*, the requisite reduction in peak width necessitates *higher efficiency per unit time*. In other words, a high α and a low H.E.T.P. are the two virtues of an ideal LC separation. But the greater of these is α.

5. Particles, Plates, and the Theory of Column Band Spreading

Though the use of plates as a measure of column efficiency has been mentioned briefly above and explained in the Appendix, let us review some concepts here so as to place them in the context of the history of LC column development.

Called *theoretical,* because it has no visible physical basis, the total *plate count* or N is a dimensionless number that may be calculated using the mathematics of statistical analysis from simple measurements made on a chromatogram. N relates the *width* of a chromatographic band to the *distance* its center [mean] has traveled through the bed. The wider the band, the more it has *spread* on its passage through the column. Band *width* or *spreading* should be minimized for greatest separation efficiency and resolution of closely eluting peaks.

The probability distribution of ten trillion molecules in a typical LC band around the band center may be described in terms of their *standard deviation, σ,* from the mean; σ is measured in the same units of distance, time, or volume as is the band width itself [illustrated in Figure 5–1].[33] A measure of the statistical dispersion of the molecules [the degree to which they are spread out] about the mean is the *variance, σ^2*; this value is not as easy to grasp in a simple mental image. If it may be assumed that the rate of change of the variance over the distance traveled by the analyte is constant, then the N may be represented by a simple formula:[34]

$$\text{Plate Count, } N = \left[\frac{V_R}{\sigma}\right]^2$$

Width [and distance] may be measured in units of volume [retention volume, V_R] or time [retention time, t_R][35] on a chromatogram. *Inside* an LC column, a band is the section occupied by the molecules of an analyte.

Martin and Synge recognized that, conceptually, after a band had passed through a column containing an infinite number of plates, a plot of the concentration of analyte molecules in that band versus elution time or volume would show a *normal probability distribution.* This is also called a *Gaussian distribution,* so named because in 1809 a mathematical genius, Carl Friedrich Gauss, defined the function that describes its shape, known familiarly as a *bell curve.* As shown in Figure 5–1, a true Gaussian curve is symmetrical about its mean and has some seemingly magical properties; the width of such a curve [w] between its two characteristic inflection points [at 0.607 × height of curve at its mean] equals 2 × sigma [σ, the standard deviation from the band center]. As noted earlier, σ^2 is the statistical variance.

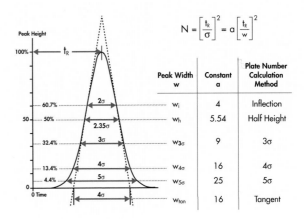

Figure 5–1: How to calculate plate number, N, from a chromatogram.[32]

To calculate N, as shown in Figure 5–1, the peak width [w] can be measured at other particular points along the height of the Gaussian curve that equal another constant multiple of σ [the letter 'a' in the next equation represents the square of this constant]. V_R equals retention time [t_R], measured by ruler or computer, multiplied by the flow rate [volume/time].

$$N = a\left[\frac{V_R}{w}\right]^2$$

The height equivalent to a theoretical plate [H.E.T.P. or H] is simply:

$$H = \text{bed length}/N$$

In 1956, inspired by their collaboration with Martin, especially on the theory and scale-up of gas chromatography, Jan Josef van Deemter and colleagues at Royal Dutch Shell, Amsterdam, set forth a mathematical equation that takes into account key factors which contribute to chromatographic *band spreading* [also referred to as *band broadening* or *dispersion*] in the beds of both GC and LC columns.[36] Shown in Figure 5–2, the simplest form of the equation [that today bears Van Deemter's name] describes H, for any given system, as the sum of the contributions of three principal processes: *A,* eddy diffusion in the spaces between particles; *B,* axial [or longitudinal] diffusion in the mobile phase; and *C,* mass transfer in the mobile and stationary phases. Figure 5–3 illustrates the mechanism of each of these three potential sources of dispersion in a packed column bed.

$$H = A + B + C$$

where:

A [interparticle eddy diffusion] $\propto d_p$

B [axial diffusion] $\propto D_M/u$ & D_S/u

C [mass transfer] $\propto (d_p)^2 u/D_M$ & u/D_S

and:

d_p = particle diameter
u = linear velocity [flow rate/area]
D_M = solute diffusion coefficient in mobile phase
D_S = solute diffusion coefficient in stationary phase

Figure 5–2: Principal terms and key variables in the van Deemter equation.

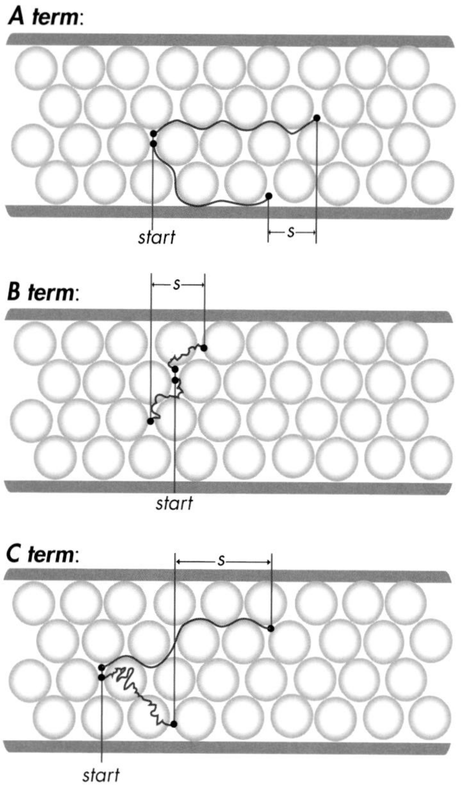

A term:

Interparticle Eddy Diffusion
Each molecule may be transported via a different fluid streamline. Local variations in fluid velocity [faster in the narrow channels] cause the molecules to travel longer or shorter tortuous paths, thereby spreading the band.

B term:

Axial Diffusion
Under static or low flow conditions, each molecule zigzags in a generally different random direction. This diffusion process may broaden the band along the longitudinal axis of the bed. In GC, this process is very important as diffusion in the gas phase is about 10^5 times faster than in the liquid phase. In liquid chromatography, for small molecules, the 'B' term becomes important at low flow velocities. Once considered negligible in HPLC, it plays a more important role in UPLC separations, emphasizing the need for higher pressure.

C term:

Mass Transfer
If the rate of equilibration of a molecule between the stationary and mobile phases is slow [slower diffusion or transfer of mass from one phase to another], then molecules attracted into the stationary phase are left behind while the mobile phase transports other molecules further down the bed, thereby broadening the band in which they are located.

Figure 5–3: Three primary mechanisms of band spreading in a packed bed. These correspond to the A, B, and C terms in the van Deemter equation [Figure 5–2], respectively. All figures show that two molecules of the same type [indicated by dots], starting at the same position along the bed axis, ultimately move a distance, s, apart due to the band spreading process indicated. Keep in mind that each band contains over a trillion molecules, each free to behave in similar fashion. Movement in a radial direction remixes a band but preserves its integrity; axial movement broadens the band. [This figure is an adaptation of a similar diagram drawn by Knox in 1978.[37]]

The van Deemter equation put the prediction of Martin and Synge on firmer footing: the highest efficiency [smallest plate height] is achieved with the smallest particle diameter. As will be seen in Chapter 10, the van Deemter equation is one of the tools that may be used to characterize the conditions for optimal operation of a particular LC system as well as to predict the requirements for achieving ultra performance.

6. The Path to High-Pressure LC

In the 1950s and 60s, gas chromatography became all the rage, and chromatographic theory developed rapidly as a result. Meanwhile, the practice of liquid chromatography remained much as it had been in the preceding two decades. Tswett had tested more than one hundred sorbents for LC suitability. Later practitioners combed cabinets and catalogs to see what kinds of materials were available from commercial sources across a wide spectrum of industries. GC columns were packed with particulate supports that ranged from ground firebrick [refractory or fireclay] to laundry detergent[38] to kieselguhr [diatomaceous earth]. In 1957, Johns-Manville, who had been selling their Celite® brand of kieselguhr to laboratories as a filter aid, began to provide a specially purified and sized version under the tradename Chromosorb specifically for use as a coated GC support.

Enterprising chromatographers also used Celite as a wide-pore, though fragile, silica support for liquid-liquid partition chromatography [LLPC].[39] M. Woelm GmbH in Eschwege was one of the first companies to respond to the need for standardized adsorbents for open-column LC separations. They created and commercialized reproducible grades of porous alumina[40] with controlled moisture content[11], surface pH, and particle size distribution. E. Merck [Darmstadt] later filled the void for standardized silicas for open column chromatography. They gave each grade a number indicating the respective average pore diameter in angstroms [40, 60, 80, 100], and ultimately made a selection of particle size ranges available. Nominally, all particle sizes with the same pore dimensions had nearly identical morphology.

These early aluminas and silicas were xerogels, formed first by precipitation from solution as cakes and then calcined or heated at high temperatures to drive out water, leaving behind an amorphous structure containing pores typically with diameters in the range of 10–1000 Å. These cakes were broken up into manageable chunks; these, in turn, were ground into smaller and smaller irregular fragments. Under a microscope, silica particles look like shards of broken glass. The smaller particles were put onto a suitable stack of fabric-like screens [woven from metal wires or plastic strands into specific mesh sizes]. These screens were arranged in a size sequence with the smallest mesh number at the top [coarse; fewest strands per inch] and the largest mesh number at the bottom [fine; most strands per inch]. The entire stack was vibrated or shaken in

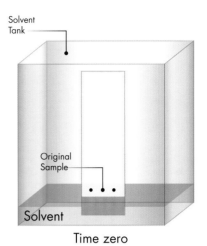
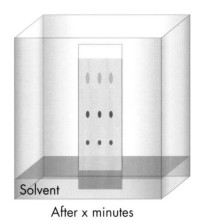
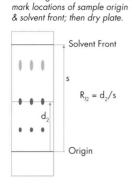

Measuring TLC Retention Factor:
*mark locations of sample origin
& solvent front; then dry plate.*

Solvent Front

$R_{f2} = d_2/s$

Origin

Time zero

After x minutes

Figure 6–1: Stahl's technique of thin-layer chromatography involves spotting a sample onto a thin layer of stationary phase silica particles affixed to the surface of a glass plate. The bottom edge of the plate is placed into a tank containing some solvent. The tank is normally kept covered to maintain a saturated equilibrium atmosphere of solvent inside. Flow is created by capillary action as the mobile-phase solvent diffuses into the dry particle layer and moves up the plate, carrying the sample with it. In the example shown, the black sample spot contains a mixture of FD&C yellow, red, and blue food dyes. Each dye compound moves a different distance relative to the solvent front, creating the visible separation. If a fluorescent salt [*e.g.*, manganese-activated zinc silicate] is added to the silica slurry before the TLC plate is made, then invisible compounds may be detected under 254-nm UV light as dark spots that locally quench the fluorescing background [non-destructive]. Alternately, a developed plate may be sprayed, after solvent evaporation, with general-purpose or highly specific reagents that produce colored reaction products [destructive]. A TLC retention factor [R_f] compares the distance the center of a spot has traveled from the origin relative to the corresponding distance that the solvent front has moved.[46]

a manner sufficient to help the smaller particles find their way through the holes in the upper screen, but not so violently as to cause excessive particle abrasion and fracture into undesirable fines or dust. Using a common open-column LC particle-size distribution as an example, particles that pass through a 100-mesh screen but are retained on a 200-mesh screen are designated as 100–200 mesh particles; the corresponding nominal particle-diameter range for this screen cut is 75–150 μm.[41] The size of larger particles was then, and continues to be reported as a range of mesh sizes. Below about 400 mesh [37 μm], screen construction and operation becomes impractical. Other techniques such as air classification and elutriation had to be developed to isolate ranges of much smaller particles. Hence, modern LC particle diameters are typically reported in micron [μm] units.

In 1950, when partition LC on paper,[42] an analytical extension of Martin and Synge's procedure in open tubes, was at its height, Prof. Egon Stahl [then at Johannes Gutenberg U., Mainz] began experimenting with a new planar LC format he named *Dünnschicht* or *thin-layer chromatography* [TLC]. Six years later he had standardized his technique [see Figure 6–1].[43] Stahl remembered that *"interest soared"* following the commercialization of the first TLC sorbent [silica gel G according to Stahl, < 400 mesh (5–40 μm, 6-nm pore diameter)—the 'G' indicated the presence of an inert gypsum binder] by E. Merck and apparatus for coating and drying a silica slurry on TLC glass plates by Desaga [Heidelberg] at ACHEMA in 1958.[44] His textbook became the bible as TLC grew to be the most widely practiced analytical LC technique in the 1960s.[45] For chemists' convenience, Merck, Analtech [Delaware], and Woelm introduced pre-coated TLC plates—thin layers [0.1–0.25 mm] for analytical, and thick layers [1–10 mm] for preparative, separations on 5 x 20 cm and 20 x 20 cm glass. Woelm and Kodak [Rochester] introduced pre-coated thick aluminum and polyester sheets, respectively; these could easily be cut to any size with a pair of scissors. Today, it is rare to find chemists who have coated their own TLC plates.

In the late 70s, following the dramatic performance increase in LC, Merck introduced miniature TLC plates pre-coated with 5–6 μm silica for HPTLC—*high-performance* thin-layer chromatography. Stahl thought this was a little ingenuous since he and his co-workers had done HP separations of nanogram amounts on plates with very thin layers and small particles nearly two decades earlier![44a] A key observation in the historical development of LC is that the use of small particles happened first in TLC, before column LC. TLC does not require a pump for mobile phase flow, and samples are not eluted from the plate. Capillary forces are strong enough to overcome gravity, and flow is sufficient for development of plates over a vertical distance of about 17–18 cm.

While TLC was replacing many paper chromatography applications,[42] a key application for partition LC, physiological amino acid analysis, was supplanted by ion-exchange column chromatography separations on an automated apparatus designed and built by postdoctoral fellow Darrel H. Spackman at the request of his mentors William H. Stein, and Stanford Moore at Rockefeller University[47] [see Figure 6–2]. They had been doing partition LC amino acid separations on columns packed with potato starch, at the suggestion of Richard Synge. Using this method, it took weeks to analyze the peptides in a protein hydrolysate. When they switched to open-column ion-exchange chromatography, they reduced the analytical time to a week. Finally, by using finely ground particles and an automated system, a complete analysis could be finished in a full day. Stein and Moore went on to use this tool in the following year to elucidate the structure of ribonuclease, the first enzyme fully sequenced. For this work they shared half of the 1972 Nobel Prize in Chemistry.[48] Spackman, being mechanically inclined, subsequently joined Spinco [Specialized Instruments Company], that in 1954 had become a division of Beckman Instruments, Inc., to commercialize the first automated amino acid analysis system.[47c]

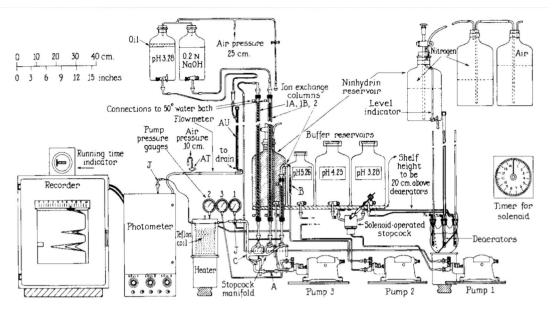

Figure 6–2: A pH sequence of citrate buffers and three columns packed with finely ground 8% cross-linked sulfonated polystyrene resin [Rohm & Haas Amberlite IR-120] were used to effect complete separation of physiological amino acids from protein hydrolysates in about 22 hours. The 50-cm-long glass column contained 25–30 μm irregular particles, while the 150-cm columns were packed with somewhat larger 40 ± 7 μm particles. Deaerated mobile phase flow rate was 500 μL/min [30 mL/hr]; back pressure was typically 35–40 psi [3 bar]. Post-column reaction, in a heated PTFE coil, of the eluting amino acids with ninhydrin was followed by photometric detection at 570 and 440 nm. A ball joint atop the column had to be opened so as to add the sample manually to the top of the packed bed. Archer Martin suggested to Stein and Moore the mechanism for accurate flow rate determination.[47b] [Figure from Reference 47b, p. 1911, used with permission.]

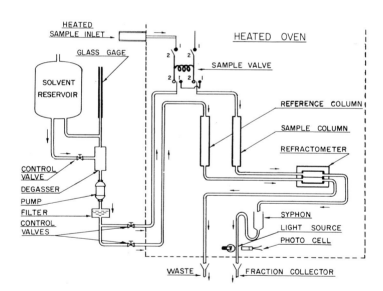

WATERS LIQUID CHROMATOGRAPHY ASSEMBLY.

Figure 6–3: This schematic diagram of Waters GPC-100 instrument fluid path was scanned from page 4 of the 1963 manual that was shipped with one of the first systems to Dr. Jack Cazes at Mobil Research.[49] Note that a prescient Jim Waters, who wrote the manual himself, named this figure "*Waters Liquid Chromatography Assembly*"! All the features of an HPLC were present in this system: valve-and-loop injector, 500-psi-[35-bar-]capable, organic-solvent-compatible pump, and optical refractive index [RI] detector. More advanced features included a siphon and photocell for accurate flow rate measurement, a means to degas the mobile phase leaving the solvent reservoir, an in-line filter, a heated oven containing both the injection valve, the 4-ft-long steel columns, and the RI cell, a strip-chart recorder [not shown in this diagram] and a fraction collector. Fluid lines were all pressure-resistant stainless steel. Up to four 4-ft long columns [7.8-mm i.d.], packed with controlled-pore-size Styragel® copolymers, were used in series. Initially, particles were screened to 140–200 mesh [75–105 μm]. Later columns used 37–75 μm particles in shorter tubes for higher efficiency.

Stein and Moore anticipated the advantages of using smaller particles at higher flow rates and pressures, but avoided attempting this avenue so as to effect a compromise between speed and the simplicity of system construction.[47b] With low pressure capability and no injection device, their integrated system did not yet possess all the key attributes of an HPLC instrument. Five years later, a system built to effect another emerging specialized technique would prove to be the first commercial HPLC.

Two months after Stein and Moore had published their paper, a young entrepreneur, James Logan Waters, founded his second company, Waters Associates, in September 1958, in Framingham, Massachusetts.[49] At the beginning of 1963, another equally brilliant Moore—John C. Moore, a chemist at the Dow Chemical facility in Freeport, Texas—called Waters to ask for a custom version of his differential refractometer. Moore wanted a tiny 10-μL flow cell with higher temperature capability. Jim Waters sent his sales manager, Larry Maley, to Texas to find out how Moore was using this special detector. After securing a license to the Dow patent for *gel-permeation chromatography* [GPC],[50] Waters and his five associates built five prototypes of the GPC-100 instrument in only six weeks, saving time by using plywood for the cabinet! In March 1963, three of these instruments were sold to Dow. Just as John Moore had been inspired by the work of Jerker Porath and Per Flodin in Uppsala on the separation of proteins by size,[51] Waters credits Stein and Moore for some of his inspiration.[49d] However, Waters also made an early commitment to materials science that would continue to set his fledgling instrument company apart from his competitors. He went to Moore's laboratory and learned how to synthesize the cross-linked polystyrene-divinylbenzene copolymer porous beads that were used as the stationary phase in the GPC columns.

While the Waters GPC instrument was quietly revolutionizing the analysis of the molecular weight distribution of synthetic organic polymers, shortening the time required from one week to a few hours, two other researchers were doing benchtop experiments with crudely constructed chromatography systems for normal-phase and partition LC. As described in fascinating autobiographical accounts of their earliest work in LC, Prof. Josef Huber in Eindhoven[52] and, independently, Csaba Horváth in Prof. Sandy Lipsky's Yale University laboratory[53] each began to experiment with smaller particle [< 50 μm] columns and higher flow rates/pressures. Huber was convinced by the work of Van Deemter that particle size was the key parameter for improving efficiency in LC and, with a student, began *"systematic investigations and exploitation of the effect of particle size in column liquid chromatography."* [see Figure 6–4]

In the fall of 1964, Horváth began his experiments directed toward approaching the high performance of GC in an LC system capable of separating non-volatile biological compounds.[54] Using a homemade system, and borrowing a method of creating a stationary phase he had developed for GC packings during his doctoral studies as the inaugural graduate student of Prof. István Halász [U. Frankfurt, 1963],[55] Horváth obtained his first HPLC chromatogram in late 1964 [see Figure 6–5]. Horváth continued to improve his prototype and encouraged the Picker Corporation to commercialize his instrument. The result was a narrow-purpose, short-lived *"nucleic acid analyzer"*.[56]

Dr. Jack Kirkland, experienced in GC, was so enthusiastic after his visit to Huber's laboratory in Eindhoven in 1964 that he became an early LC adopter [and ultimately, with Lloyd Snyder, a modern LC mentor to a generation of practitioners[57]]. Kirkland returned to his laboratory in Wilmington, Delaware, and persuaded his manager to permit experiments on this nascent technique.[58] He succeeded in building an LC system, converting an online UV process analyzer

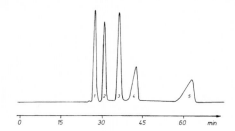

Figure 6–4: Josef Huber tentatively called this the first HPLC chromatogram. It was published in the January 1964 M.S. thesis of his student J.H. Quaadgras [U. of Technology, Eindhoven]. A 5 x 770 mm glass column was packed with 37–50 μm liquid-stationary-phase-coated diatomite particles. At a flow velocity of 0.5 mm/sec and a pressure < 200 psi [14 bar], a simple mixture of test analytes was separated in a little more than one hour. Efficiency measured on peak 3 [2-aminobenzoic acid methyl ester] was 4000 theoretical plates.[52] [Reproduced with permission from Reference 52.]

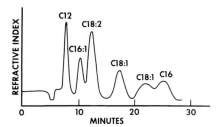

Figure 6–5: One of Horváth's earliest chromatograms from late 1964, obtained on a homemade assemblage of modular components. A one-meter-long, 1-mm-i.d. column was packed with 100–200 mesh [75–150 μm] glass beads coated with a pellicular layer of graphitized carbon black. Ionized fatty acids in an ethanol mobile phase containing 10^{-4} M. sodium hydroxide were separated and then detected by refractive index. In 1965, Horváth's second-generation LC was capable of operating at pressures up to 1000 psi [70 bar]. [Reproduced with permission from reference 53a.]

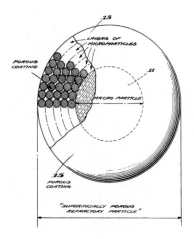

Figure 6–6: Schematic of a pellicular packing shown in Kirkland's 1970 patent. It claims macro particle cores from 5–500 μm and a layer(s) of 0.005–1 μm microparticles that constitutes from 0.002 to 25% of the total particle volume. The process of making such particles was developed by Kirkland's DuPont colleague and silica wizard, Ralph Iler.[59]

into a sensitive detector.[60] This success encouraged DuPont to enter the LC instrument business in the late 1960s. Simultaneously, other companies began to commercialize components and LC systems. Waters, for example, introduced in 1967 the ALC-100, a system designed specifically for analytical liquid chromatography at > 1000 psi [70 bar] with 25–40 μm particles.

One of the earliest improvements in the efficiency of LC packing materials was the development of what Csaba Horváth termed *pellicular* phases, *i.e.*, solid-glass-core particles with a thin outer coating or *pellicule* of a porous solid phase.[61] Horváth's mentor in Frankfurt, Prof. István Halász, became a consultant for Waters Associates and encouraged their development of the first commercial pellicular silica packings [tradename: Corasil]. These 37–75 μm particles, packed in 2-mm-i.d., two- and three-foot long columns, afforded a dramatic advance in column efficiency.[49 a,b] Nearly a month later, at the end of January, 1970, the second commercial silica pellicular packing , Zipax™, was introduced [see Figure 6–6]. A third such packing from E. Merck, whose name, Perisorb, incorporates the Latin prefix *peri* meaning around or surrounding, debuted in late December 1971. With their thin shells and controlled surface porosity, these pellicular particles, while very limited in capacity and surface area, equilibrated quickly, and analytes could not diffuse very deeply into a 1–10-μm thick porous layer. Pellicular packings exhibited high efficiencies at a time when commercially available materials and convenient packing techniques did not yet exist for making columns using totally porous particles whose diameters were similar to the thickness of the pellicule layers.

At the Pittsburgh Conference in March 1970, Horváth gave a lecture in which he referred to his methodology as *HPLC*, his shorthand for *high-pressure* or *high-performance liquid chromatography*.[62] Within the hour, he was amazed to see that vendors in the exhibition hall had erected hastily handwritten signs stating, "We do HPLC!"[63] While others came before him, Horváth is often credited as the father of HPLC;[64] a catchy acronym, an engaging personality, and a bully academic pulpit can have a powerful effect on history!

It was Jim Waters' commitment to helping his customers be successful that led him in 1970 to pack some columns, take an ALC-100 instrument into the Harvard University laboratory of Prof. Robert Woodward, and solve some key separation problems involving the purification of milligram amounts of isomeric intermediates on the pathway to the first total synthesis of Vitamin B_{12}.[49] When he, at the urging of his peers, finally summarized his success, Woodward wrote: *"Here I should say that of absolutely crucial importance to all of our further work has been our taking up the use of high pressure liquid or liquid-liquid chromatography to effect the very difficult separations with which we were faced from this point onwards. The power of these high pressure liquid chromatographic methods hardly can be imagined by the chemist who has not had experience with them; they represent relatively simple instrumentation, and I am certain that they will be indispensable in the laboratory of every organic chemist in the very near future."*[65]

This success energized HPLC out of its long induction period and initiated a steep slope in its growth curve as organic chemists everywhere were eager to heed Nobelist Woodward's admonition. Soon a single peak in HPLC supplanted a single TLC spot as a criterion for purity in synthetic chemistry.[65] HPLC had arrived!

7. Modern HPLC Watershed

Josef Huber had taken up the LC flag at the Zlatkis meeting[66] in January 1969, and waved it strongly. He reexamined the general theory of elution chromatography and championed the potential of LC in the face of the then predominant GC and TLC techniques. In a section of his published lecture entitled, *"Speed of Liquid Chromatography,"* he stated succinctly, *"The column is the heart of the chromatograph."*[62] Though as a literal biophysical analogy this may not have been entirely accurate [a heart is a pump, after all], Huber wanted to emphasize the central role the column plays in creating an efficient separation as quickly as possible.

Figure 6–7: One of the pioneering LC separations Jim Waters did in Woodward's Harvard U. laboratory, using normal-phase LC on 37–75 μm pellicular silica [in a bank of five 2-ft-long, 2-mm-i.d. columns connected in series] to separate and isolate two structurally complex isomers [differing only in the orientation of the two substituents at position 13, highlighted above in yellow], one of which was the desired intermediate in the total synthesis of Vitamin B_{12}. Baseline separation was achieved by using a manual valve to *recycle* the overlapped peaks after they eluted back through the columns. In this way, column efficiency was effectively increased in an additive fashion while sample capacity was also maximized.

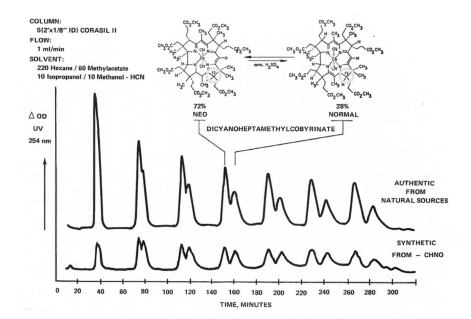

Figure 7–1: Since his first HPLC experiments, Huber continued his zealous mission to reduce the particle size of packings for maximum efficiency and minimum separation time. This chromatogram, first displayed at a Column Chromatography meeting in Lausanne in October 1969, illustrates a major improvement in a test mixture separation by using a 2.8 x 230 mm glass column packed with a liquid-coated diatomite phase having a particle size of 5–15 µm! [Amazingly, he was able to pack this small-particle column successfully using a dry tamping technique; see Chapter 8.] Compare the efficiency and separation time in this chromatogram with that in Figure 6–4; the analytes represented by peaks one and two are the same in both. [See reference 52 for details; figure reproduced with permission.]

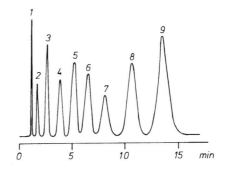

In the same lecture, Huber listed eight primary capability requirements for an LC system—as valid today as they were in 1969!:

- *High speed of separation*

- *Repetitive separations on the same column*

- *Real time detection*

- *High precision of the quantitative analysis*

- *High precision of the retention data*

- *Suitable for automation*

- *High production rate for preparative application*

- *In-line combination with other methods*

Confirming the earlier postulates of Martin, Synge, and Van Deemter, he concluded from his mathematical analysis [using a mass-balance rather than a statistical approach] that LC could approximate the speed of packed-column GC. To achieve *"a better separation in a shorter time"* would require *"the use of small uniform particles and the preparation of a regular column packing. The price which has to be paid is the high pressure drop."*[52]

Despite strong resistance from naysayers who argued that particles with diameters smaller than 20 µm would be unstable and agglomerate, or that packing them into homogeneous beds would require *"black magic,"* Huber persisted in his experiments, and 10 months later he showed the result seen here in Figure 7–1.

Huber spent a sabbatical year in Boston in 1973 where he met Jim Waters; sharing much in common, they became great friends [see Figure 7–2]. No doubt their bond was strengthened as Huber witnessed firsthand the commercial debut of three major, co-enabling breakthroughs in HPLC between December 1972 and mid-1974. A Waters chemist, Dr. Richard Vivilecchia, and his colleagues developed the first commercially available HPLC columns packed with 10-µm irregular-silica particles and the first 10-µm *monofunctionally* bonded octadecyl-silica reversed-phase sorbents[67] [µPorasil™ and µBondapak® brands respectively]. Waters engineers Burleigh Hutchins and Louis Abrahams designed the first dual-reciprocating-piston HPLC pump capable of smooth solvent delivery at 6000 psi [see Figures 7–3 and 7–5], and the first septum-less injection system also designed for 6000-psi [400-bar] operation [see Figure 7–4]. The dramatic success of these core products not only drove the ensuing rapid growth of Waters Associates; it also hastened the creation of a competitive industry that grew exponentially as the utility of HPLC spread to every corner of chemical and biochemical endeavor.

Figure 7–2: This publicity photo was staged in 1973 at Harvard U. in the laboratory of Prof. Robert Woodward, at right [with notorious cigarette in hand!]. Jim Waters, center, had worked with postdoctoral fellow Dr. Helmut Hamberger, left [autograph on right sleeve] three years earlier on critical separations like that shown in Figure 6–7. The Waters ALC-100 system is at the right; a bank of 3-ft-long Corasil columns [37–50 µm, 2 mm i.d.] stands vertically in the column bay at the far left side of the instrument. Prof. Josef Huber, second from left, was in Boston that year. Huber had been a chromatography consultant to Woodward's Swiss colleague, Prof. Albert Eschenmoser [ETH, Zürich], in the total synthesis of Vitamin B_{12}.

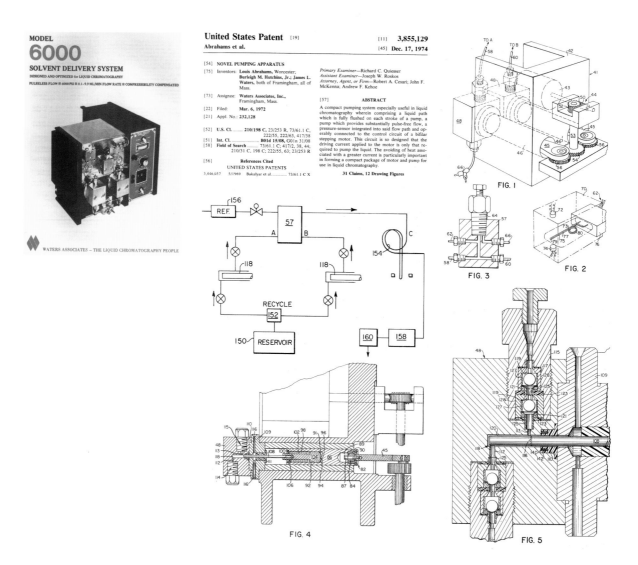

Figure 7–3: Before the novel Model 6000 pumping apparatus was created, HPLC systems used either gas-pressurized fluid reservoirs or dual-piston chemical metering pumps to move mobile phase through columns. The former were pulseless, but their fixed capacity limited the total duration of HPLC experiments, and safety was an important concern. The latter were not able to deliver small volumes of organic liquids with the required accuracy and precision, and significant pressure pulses were created each time flow switched between cylinders. Both were limited in pressure capability. By contrast, the M6000, announced in December 1972, [characterized in the patent abstract, by what proved to be a huge understatement, as *"especially useful in liquid chromatography"*] had a chemically inert fluid path that was fully flushed on each pump stroke [see schematic above], non-circular gears [see illustration labeled FIG. 1] that minimized pressure pulses at cylinder crossover to provide substantially pulse-free flow [~ 1% variation], and a pressure sensor [using a Bourdon-tube and photoelectric detector, FIG. 2] that limited the current to the bifilar stepping motor to only those times when liquid had to be pumped. This also provided a simple means of compensating for the compressibility of the fluid being pumped, as well as that of the pumping system itself. By avoiding heat associated with higher currents, the need for bulky heat dissipation devices was obviated, thereby enabling the overall pump package to be as small as possible. Small-diameter, inert, sapphire pistons, innovative ruby-ball-and-seat check valves [FIG. 5] in each head, and inert, spring-loaded seals enabled flow delivery at pressures up to 6000 psi [400 bar]. A mixing block [FIG. 3] unified the output of each piston into a single high-pressure flow stream, and its integral valve permitted the stream to be diverted, as required, to flush the reference cell of a differential-refractive-index detector. Mechanical links isolated the pistons from the gears [FIG. 4]; the gear train was immersed in a sealed oil bath for constant lubrication.[68] In the photo above on the cover of the first product brochure, note that the dark narrow box at the right, bearing the face plate with switches and dials, isolated the sophisticated electronic control circuitry from the fluid delivery mechanics in the left section. The entire shell was a single cast block of aluminum with appropriate cutouts and holes drilled precisely by automated machining systems. Many of these rugged pumps were used for decades of HPLC experiments; some were used in sophisticated non-LC metering applications in operations requiring reproducibility and chemical compatibility. More than a few are still in operation today! The basic design of the M6000 and subsequent models lent itself readily to modifications by various researchers to accommodate lower and/or higher flow-rate and pressure capability. For example, in 1999, Douglas Wittmer at Waters modified a later generation of these pumps [Model 515] to operate at 20,000 psi [1400 bar]. These were used in the earliest experiments to develop the initial UPLC columns and system components.[69]

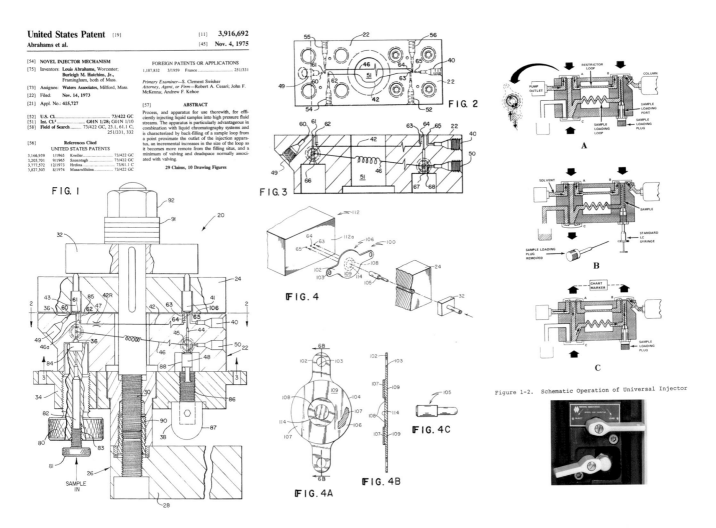

Figure 7–4a: Unless you had endured the leaks, bent fixed needles on expensive syringes, and sample losses characteristic of the learning curve for practicing the art of making on-column injections through an elastomeric septum against high back pressure, you could not fully appreciate the excitement in the LC community caused by the announcement of the Waters Model U6K septum-less injector [U = universal; 6K = 6000 psi [400 bar] capability] in late 1973. At that time, there was no commercial apparatus *"suitable for use in injecting small liquid samples into high pressure systems without undesirable risk of loss, contamination, or dilution [band spreading] of the sample and also loss of constant pressure in the system."* Unique features included: solid-state plumbing [tiny fluid passages drilled into metal manifold blocks [see cross-sections in FIG. 1, FIG. 2, and FIG. 3 above]; a novel, zero-volume, flexing metal-diaphragm valve [FIG. 4 and detail in FIG. 4A, FIG. 4B, and FIG. 4C]; a selection of fixed-volume sample loops made of small-bore stainless steel tubing [#46 in FIG. 3] that were back-filled at atmospheric pressure with a hypodermic syringe; and a parallel flow-resistance tube [#42 in FIG. 3] that continues constant flow and pressure in the downstream column and permits a make-before-break switch of the fluid stream to the sample loop during injection [schematics in Figure 1–2 above, reproduced from an early product manual].[70]

Figure 7–4b: Not quite four years later, further improvements [wiping seal; drip-proof exterior needle wash] to the core U6K technology became the basis for the critical mechanism in the first automated HPLC injection system for unattended processing of multiple samples [Model 710; tradename WISP (Waters Intelligent Sample Processor)].[71] Notably, this instrument was the first instance of the use of embedded microprocessors [Intel 8080] in modern LC instrumentation.

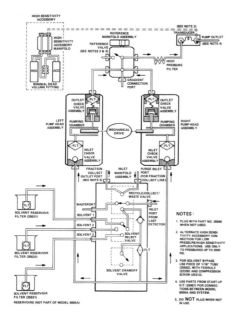

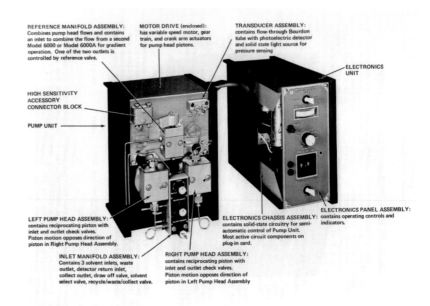

Figure 6-1. Hydraulic Components, Circuit Diagram

Figure 1-2. Major Components of Model M6000A

Figure 7–5: As with any worthy pioneering product, hundreds of improvements were made to the M6000 pump over its lifetime.[72] As shown above in two illustrations from its instruction manual [October 1975], the Model 6000A featured a new inlet manifold assembly that contained 3 valves. The late Lawrence Finn extended some of the valve concepts embodied in the U6K injector to create a sandwich of three stainless steel blocks separated by inert fluorocarbon gaskets. A plethora of tubing and fittings were obviated by machining tiny channels and holes into the interior surfaces of the blocks to form the fluid conduits that a user connected alternately by rotating valve knobs. Small, totally swept volumes were critical to minimizing both band spreading in the recycle mode and solvent carryover when switching mobile phases. As yet another example of seminal technology, this compact, simple, yet highly sophisticated, solid-state plumbing design presaged modern miniaturized fluid circuits.

8. A Cascade of Smaller Particles

As so often happens in the early days of an exciting technical arena like the development of HPLC stationary phases and instrument systems, pioneering, international research and development efforts occur concurrently and independently. This particular arena was unique in that much of the work was done under corporate, rather than academic, aegis. Exactly who started which experiments when is virtually hidden in proprietary annals and fond, though foggy, memories. However, trademark applications recording first commercial shipments of products, if accurate, may establish a chronological basis that competitive minds view as grounds for bragging rights. Despite the fervor of those racing to establish a market position, what ultimately matters, of course, is that diverse and interesting scientific approaches to solving important technical problems lead to a better understanding of materials and methods that forms a foundation for the future of the discipline, the income to fund further research and development, and the creation of improved tools that enable discovery and the advancement of science.

Apart from better pumps, injectors, and detectors, there were three fundamental improvements that enabled HPLC to become the prevailing analytical technique.

First: The creation of suitable particles with diameters ≤ 10 μm.

Second: The refinement of slurry-packing methods to enable the formation of stable, high-efficiency column beds from these small particles.

Third: The application of these new HPLC columns to a wide variety of important samples in seminal areas of research and development.

Of these, the second, slurry packing, was singularly significant. To gauge the enormity of its impact, you need only refer back to Figure 7–1 and pity the poor efficiency Huber was able to coax out of 5–15 μm particles in a 23-cm long dry-packed column!

Microparticle pioneers

Let us review first the development of small-particle stationary phases for HPLC. While a thorough review of this field is beyond the scope of our book, we are fortunately still in a position to relate some firsthand history of the seminal efforts of a few pioneers.[73] As mentioned above, Dick Vivilecchia introduced 10-μm silica columns shortly after the December 1972 announcement of the M6000 pump. Being an analytical—not a synthetic—chemist, he purchased from another manufacturer a suitable irregular silica, then modified its surface and morphology to suit his design. Within three months, he had synthesized μBondapak C_{18} packings using the same silica substrate.[74] Soon thereafter, 10-μm irregular silicas and corresponding bonded phases became available in packed columns from other vendors: e.g., LiChrosorb® from E. Merck and Partisil from Whatman.

While small-particle irregular packings were an obvious extension of the traditional open-column LC phases, it was thought that spherical particles might enable HPLC columns to have a more stable and uniform packed-bed structure with lower flow resistance. One of the first to make porous spheres of metal oxides was Dr. Michael Holdoway. As he relates, "In the 1960s, I was part of a group in the Chemistry Division, Atomic Energy Research Establishment, Harwell, UK, investigating the use of a sol-gel process to prepare spherical particles of uranium and plutonium oxides for use as fast reactor fuels. The process consisted of preparing colloidal suspensions of the oxides, forming droplets in an immiscible solvent, and removing water to convert the liquid sol to solid gel spheres. In 1965, Parliament passed the 'Science & Technology Act'. One of the results of this was that Harwell was encouraged to apply processes and methods that were developed for the Atomic Energy Industry to the non-nuclear field and find outlets for them in British industry. I was one of the people designated to do this, and I decided to research colloidal alumina as a likely candidate. One of the offshoots of this activity was the use of the material as a wash-coat for exhaust gas catalysts systems that, I believe, is still used today to increase the adhesiveness of the platinum catalyst coating."

"About 1970, a letter from Prof. John Knox, of Edinburgh University, was passed to me. In it he outlined the (then) new field of HPLC and, having learned of our experience in spheroidisation of metal oxides, inquired whether we could prepare spherical silica. I responded that we had no experience in silica but that I could meet his specification with alumina. I duly sent him a sample of 5-micron spherical porous alumina."

"Prof. Knox had done some prior work using solid-glass spheres, matching a theoretically predicted [van Deemter efficiency] curve, but, on testing the alumina sample, he achieved identical results for the first time in a practical system. However, although excited by the result, he insisted that silica was the material of choice. So, in conjunction with a fellow worker, Mr. Anthony C. Fox, we investigated the conversion of a silica sol [which was, and still is, commercially available] into spherical porous silica. This later became 'Spherisorb', but, since a government establishment was not allowed to enter into commercial activities, it was necessary to find an industrial partner [Phase Separations] and receive a royalty on their sales. For about 5 years, the material was manufactured at Harwell and supplied to Phase Separations for sale. In 1978, they exercised their contractual option of taking over the manufacture and the trade name themselves. When they did this, I resigned from Harwell and joined them as Technical Director with responsibility for Spherisorb® production."[75] The initial product was a 5-μm spherical silica; the line was extended to 10 and 20 μm. About 1980, the first 3-μm particles were introduced at Pittcon®.

Larger particles of porous spherical silica had in fact been made by a sol-gel process in Antony, France at the Péchiney-Saint-Gobain research center about 1965.[76] Patents were issued in 1967, the same year that these packings, named Spherosil, were sold in France and distributed in the United States by Waters Associates using their Porasil trademark.[77] Sold in six grades, differing by pore size distributions centered from less than 100 to more than 1500 Å, Porasil packings were first used for GPC separations.[78] Some Porasil materials were coated by users with traditional liquids such as SE-30 for GC or LLC partition separations. Smaller particle sizes [37–75 μm] were packed into 2-mm-i.d. and 7.8-mm-i.d. columns 2–3 feet long for analytical and semi-preparative LC, respectively. In mid-1975, Guillemin and his colleagues reported making and testing some experimental 5–10 μm Spherosil packings with excessively high surface areas [860–1100 m²/g] and extremely small pore diameters [20–35 Å] for normal-phase and partition LC applications.[79] With such non-ideal characteristics,[80] these packings did not prove to be commercially viable.

As a result of the collaboration with Prof. István Halász, Waters sold early in 1969 under the trade name Durapak the first commercial chemically bonded, bristle [or brush] phases.[81] Entities such as Carbowax 400 [polyethylene glycol, MW 400] or 3-hydroxyproprionitrile [OPN] were bound to surface silanols on Porasil, as well as the pellicular Corasil packing, using an ether linkage. Covering pellicular or totally porous silicas with a coated or bonded polymeric layer was the first attempt to mimic an LLC packing without the drawback of the mechanical instability and subsequent bleeding of the liquid coating that compromised column and detector performance in HPLC. Ultimately, covalently bonded brush phases proved superior to polymeric phases in efficiency, presumably due to better mass transfer properties.[81] Dick Vivilecchia made a further improvement by using his aforementioned new monofunctional octadecylsilane to form a siloxane link to the surface in Bondapak C₁₈/Porasil B and Bondapak C₁₈/Corasil in late 1973.[67] Siloxanes are significantly more hydrolytically stable than silyl ethers [see Figure 8–1].

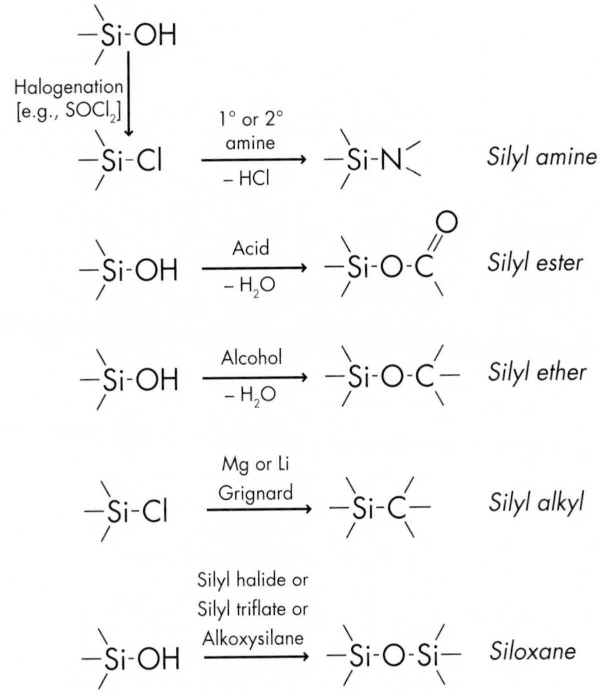

Figure 8–1: In the 1970s, many synthetic routes toward chemically bonded silica phases were tested. A silyl-ether bond [third from top, above] was the link used in the first commercial brush phases [Durapak]. This type of bond is readily hydrolyzed in aqueous or alcoholic media, thereby limiting the utility of such phases. [NOTE: In the literature, this type of bond has erroneously been called an ester.[82]] Even less hydrolytically stable and not practical for bonded phases are silyl esters [second from top]. Silyl amines [top] are a bit more stable, but limited to a pH range of 4–8.[83] These never proved commercially viable, nor did phases in which a halogenated silica surface was reacted to form a silicon-carbon alkyl bond directly [fourth from top].[84] It was difficult to completely react the surface and to wash out the residual metal salts. The most successful surface bonding approach proved to be the formation of a siloxane bond [bottom]. Silanes containing one, two, or three reactive groups were used for mono-, di-, and tri-functional bonding, respectively. Each approach has its advantages and disadvantages. For example, monofunctional bonding may yield a more reproducible surface coverage, but the single point of attachment may lead to easier hydrolysis of the bonded ligand. A trifunctional ligand may attach in 1–2 places on the surface, and/or polymerize by reacting with adjacent ligands, thereby requiring more bonds to be hydrolyzed before the ligand can be removed from the surface. However, the bonding reaction is harder to control reproducibly, and care must be taken to hydrolyze all remaining reactive moieties [e.g., silyl-chloride functions] and achieve complete reaction. You can learn much more in several books and reviews that deal with the chemistry of silica and bonding reactions.[85]

Probably one of the most prolific inventors of LC phases is Prof. Klaus Unger, U. Mainz. His research was funded, in large part, by E. Merck, Darmstadt, to which company, in return, he assigned many of his patents. Holdoway and Unger were unknowingly devising pathways to spherical silica at about the same time.[86] Merck launched LiChrospher in late November 1973, about 7 months after the introduction of Spherisorb. Unger's research program continued to innovate through the next three decades, even into his retirement. In 1979, he published his classic compilation of research on silica for LC.[87] Most recently, he has reviewed the highlights of five decades of HPLC column development.[88]

At the Physical Chemistry Institute, U. Saarbrücken, in 1972, Imrich Sebastian and Prof. István Halász, having benefited by consulting Unger on a key issue, developed yet another kilogram-scale process for making porous spheroidal silica, using ion exchange first to remove cation and acid-anion impurities in a polysilicate solution before emulsifying and coagulating the treated solution in a water-immiscible organic solvent to form particles.[89] Uwe Neue remembers that in this *"spray-drying process ... pores were created by adding a water-soluble polymer that could be washed out (first attempts) or burned out (final process). I don't recall if the polymer was dextran or polyethylene oxide."*[90] In late 1976, Sebastian joined the Macherey-Nagel company in Düren, which had licensed his process, to commercialize Nucleosil silica in 1977.

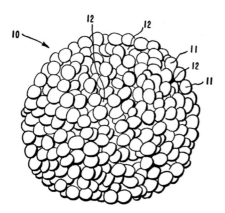

Figure 8–2: Among the many products and processes invented by pioneering silica chemist Ralph Iler[91] during his tenure at E.I. du Pont de Nemours and Company, Wilmington, Delaware, were aqueous dispersions of colloidal silica [particle diameter: 5–50 nm; sold under the trade name Ludox].[92] More than a decade later, he mixed such a sol with melamine and formaldehyde, copolymerized the monomers, thereby coacervating the organic material into microparticles containing the silica colloidal spheres. Then, by burning away the organic residue at about 550 °C, what remained was a powder of more or less spherical silica particles consisting of an interconnected array of the conjoined colloidal particles separated by interstitial passages of porosity. Subsequent calcination at about 900–1000 °C was done to further strengthen the silica particles; this treatment also reduced their microporosity and surface area.[93] Shown above is the illustration that appears in both the Iler patents[93] and a third patent by Kirkland, that describes the utility of such particles for HPLC.[94] In the drawing, #10 refers to the entire microparticle [0.5 to 20 µm in diameter], #11 denotes the individual colloidal silica particles, and #12 indicates the pore network created by the spaces between the silica spheres after the organic polymer has been vaporized by total oxidation. Kirkland published a preliminary report on HPLC using experimental batches of these particles in October 1972[95]. DuPont Zorbax columns, containing 8–9 µm silica, were first sold in February 1974; bonded phases and a 5 µm version followed not long after.

Yet another process for making spherical silica was devised by Paul Raven[96] and Prof. John Knox at the Wolfson Liquid Chromatography Unit[97] of the Department of Chemistry, University of Edinburgh, in 1974. A year later a spate of chemically bonded packing materials were synthesized on both Holdoway's alumina [20 µm] and the Wolfson silica particles [as small as 6 µm].[98] By June 1976, Shandon had adopted the processes, and in January 1977 they first sold the Hypersil spherical silica and Hypersphere family of bonded phases.

Success with slurry packing was critical

As mentioned earlier in this chapter, HPLC would not have been possible without successful methods for packing small-particle columns. Larger silica or alumina particles [> 25–30 µm] were traditionally packed dry into suitable tubes. Vibration was the easiest approach to settling the bed, but, if done too vigorously, it would unsettle portions of the bed that were already packed, leading to poor performance. One of the best procedures for dry packing was the RTP, or rotate-tap-pour method.[99] Larger particles, especially those with a solid core, are heavier and settle readily with minimal assistance. As Sie and Van den Hoed observed when devising complementary slurry-packing techniques, even if you constantly stir a slurry of larger particles in a fluid so as to keep the particles in suspension and the density of the slurry uniform, as soon as you begin to pour the slurry into a column tube, the larger, heavier particles will begin to sediment by gravity.[99]

It is a much different situation once the diameter is reduced to ≤ 10 µm. A mass of such smaller, lighter particles behaves more like a fluid than a pile of rocks or marbles. Light particles tend to float in dense fluids. At Waters Associates in the 1960s, a technique used successfully to pack light, though large, polystyrene-polymer [Styragel] particles into GPC columns involved creating a slurry in which the density of the polymer and that of the fluid were matched, or balanced. Early HPLC practitioners adopted the balanced-density slurry technique to pack efficient columns with 37–44-µm pellicular silica,[100] 5–6-µm spherical silica,[95] 5–10-µm irregular TLC silica,[101] and 4–8-µm spherical silica particles.[102] Variations on the theme such as using high-density solvents for the slurries, filling tubes under constant flow rate or constant pressure were also tried with some success. Ultimately, all slurry-packing methods are actually a form of high-pressure filtration. As Neue points out in his excellent review of column packing techniques, none of these early methods were suitable for commercial column production, so column manufacturers ultimately devised and refined their own proprietary procedures.[103]

It is one thing to create small batches of porous silica particles in the lab, but quite another to make them reproducibly, and even more difficult to scale up the synthesis, sizing, surface treatments, washing, drying, and packing processes that must yield annually many thousands of columns, each of which contains 10–15 billion particles of similar size and morphology that collectively perform HPLC separations identically batch to batch and lot to lot.[104] As Unger notes, it took about two decades for commercial HPLC columns, most of which were packed by empirically derived, proprietary processes, to achieve the level of reproducible performance desired by the LC community.[88] This was especially true for 3 µm particles that began to be offered after 1980. They had to be packed in much shorter tubes, and it required as much art as science to obtain reasonable bed stability and optimum efficiency. Shorter [and/or smaller diameter] tubes meant smaller column volumes, and, in order to gain the efficiency benefit of smaller particle diameters, corresponding modifications had to be made to LC systems to reduce volume and extra-column band spreading.[105]

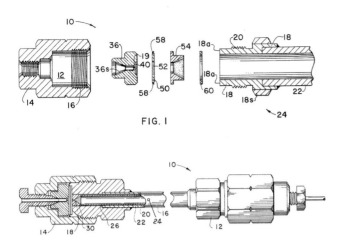

FIG. 1

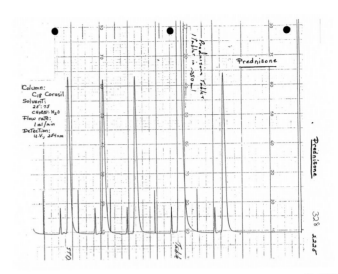

Figure 8–3: One of the keys to successful performance of small-particle HPLC packing materials was the design of the column into which they were packed. In one example shown on top in FIG. 1 above, Louis Abrahams and Manuel Russo devised novel, second-generation end-fitting assemblies that met several goals: (1) dead space or unswept volume was virtually eliminated so as to minimize band spreading and optimize sample distribution; (2) chemically resistant materials of construction were used for all parts in contact with fluid while less resistant but more mechanically desirable materials could be used for the connector parts subjected to wear; (3) fittings could be tightened to operate at 6000 psi [400 bar] without damage to the structure of the column; and (4) wearing parts such as tubing and ferrules could be replaced with minimal disturbance of the packed bed and other functional end-fitting components such as cones and filters.[106]

A second novel aspect, shown in the lower diagram above, was the patented tube-within-a-tube used for the body of the column.[107] An inner sleeve of chemically resistant stainless steel with a very smooth, mirror-like interior surface was expanded to its elastic limit and contained within an outer reinforcing tube with a yield strength of over 100,000 psi [6900 bar]. So rigid was this outer tube that its diameter would increase less than 0.01% when subjected to an internal stress of 5000 psi [350 bar]! While most single-wall HPLC column tubes at the time had an i.d. of 4.6 mm, Abrahams' unique design dropped the i.d. to 3.9 mm. This smaller diameter reduced the analytical column volume by 28%, thereby lowering mobile phase consumption by the same percentage. It also offered the potential to reduce sample size and/or increase band concentration for higher sensitivity.

A pathfinder leads the way

As mentioned at the start of this chapter, the third key element in the rapid dissemination of HPLC was the promotion of key applications in important fields. Pharmaceutical industry veteran Charles Pidacks stepped forward, with his enthusiastic and engaging personality, as a champion to convince his former peers of the utility and value of this nascent technique.[39] After joining Waters Associates in 1972, he achieved separations of every commercially available pharmaceutical compound formulation and/or preparation, first on Bondapak C_{18}/Corasil and then on μBondapak C_{18} columns. He shared this compendium [his *Pharmaceutical Notebook*, see Figure 8–4 at top right with every pharmaceutical chemist/chromatographer in the industry—and they all knew him on a first-name basis! He was a fixture at meetings such as the Land O'Lakes Conference. He used to end his presentations with a cartoon drawn by James Clifford showing a well-used column flying with angels' wings, insisting that when a column died, as it inevitably must, it goes, not to hell, but to heaven![108]

Figure 8–4: Charles 'Charlie' Pidacks kept his finger on the pulse of the pharmaceutical industry.[39] His proselytization was key to the rapid acceptance of HPLC as the preferred analytical technique in this all-important application arena. [1983 photo by Dr. Yuri Tuvim]

Now you have been introduced to some of the HPLC pioneers whose insight and innovation provided the stiff breeze that gave Huber's HPLC flag a permanent wave. If you wish at this time to take a detour off our track tracing the origins of UPLC technology, we recommend that you venture into Dr. Uwe Neue's comprehensive and definitive treatise to learn much more about the intricacies of creating, evaluating, and using HPLC columns.[27d] However, we must mention a few more ideas that, in hindsight, set the stage for the next plateau.

From high purity to hybrid

Both Unger and Knox realized that bonded-silica reversed-phase HPLC packings, as useful as they had been proven to be, had some limitations. The siloxane bonds that formed the silica backbone, as well as the links between the organofunctional groups and accessible surface silanols, would hydrolyze, especially at pH extremes. Further, it was impossible to completely mask all the silanol activity via bonding and end capping. Covering particles with a polymeric layer was one option, but these phases suffered from poor mass-transfer properties that reduced their efficiency. Carbon seemed to be a possibility, being inert, hydrophobic, and biocompatible, but a satisfactory, mechanically stable, porous carbon particle needed to be invented. Unger got there first, but his solution, based upon activated carbon or coke, never became commercially viable.[109] Knox, on the other hand, managed to make a successful product by coating a template particle such as silica with a monomer mixture, then inducing polymerization, graphitizing the organic polymer by heating in an inert atmosphere, and finally dissolving away the inorganic template. Shandon sold the product as Hypercarb, and it remains available today.[110]

Apart from their disadvantageous friability, carbon packings proved difficult to functionalize by conventional large-scale wet-chemical means, so their surface could not easily be modified in ways that might provide useful alternatives when seeking optimal separation selectivity. It had occurred to Unger that carbon functionality might be incorporated into a modified silicon dioxide by direct synthesis. He had pioneered the synthesis of silica from the controlled hydrolysis and polymerization of tetraalkoxysilanes [see Figure 8–5].[86] Why not include an organotrialkoxysilane in the reaction mixture to produce a hybrid co-polymer with significant organic functionality throughout the silica skeleton?[111] Unfortunately, especially if the organic moiety contained too many atoms [e.g., octadecyl], the morphology and stability of the particles obtained by Unger's process were not optimal for HPLC. This good idea lay fallow for nearly a generation [see Figure 8–6].

High-Purity Silica Manufacturing Process

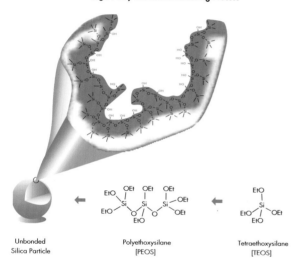

Unbonded Silica Particle   Polyethoxysilane [PEOS]   Tetraethoxysilane [TEOS]

Figure 8–5: In the 1980s, coincident with the rise in applications of HPLC for the analysis of biomolecules and basic pharmaceutical compounds, residual metal ions in the silica substrates for LC packings became a concern. Ultimately, the source of these metals, including sodium, aluminum, calcium, magnesium, and iron, could be traced to the solutions of mineral silicates used to create precursors of the xerogels or sols [and perhaps to the processing equipment with which the silica came into contact].[112]

Inertsil from GL Sciences in Japan, introduced in 1986, was the first in a succession of commercial LC small-particle, high-purity, spherical silica packings. This new generation of particles was made directly by polymerization and hydrolysis of pure tetraalkoxysilanes. Each manufacturer created a proprietary variation on Unger's theme. In the figure above, we show the generic scheme used to synthesize the silica used in Waters Symmetry® packings. Though not the first such product [1994], the extraordinarily tight controls over its manufacturing process made Symmetry silica the ultimate standard in reproducibility,[113] a benchmark by which the new generation of sophisticated HPLC mobile-phase delivery systems, introduced the same year, were measured.[114]

Learning from the experience of making Symmetry silica, in 1998, Dr. Zhiping Jiang and his colleagues invented a process to incorporate a Si—CH₃ moiety throughout the backbone of a spherical silica particle with a pore structure and particle morphology optimized for HPLC [see Figure 8–6].[115] The first products of this research were the XTerra® brand packings introduced at HPLC '99 in Granada, Spain. These inorganic-organic hybrids became the new paradigm in LC packings, bringing Unger's original notions to fruition. They proved to be a prelude to a new generation of hybrids that would catalyze the creation of UPLC technology.

High-Purity Hybrid Silica Manufacturing Process

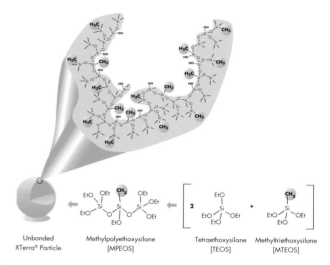

Unbonded XTerra® Particle   Methylpolyethoxysilane [MPEOS]   Tetraethoxysilane [TEOS]   Methyltriethoxysilane [MTEOS]

Figure 8–6: Since one-third of the monomer mixture is methyltriethoxysilane [MTEOS], there is a corresponding reduction of the total population of surface silanols in this hybrid particle [compare this illustration with the structure in Figure 8–5]. Yet, in the accessible pores and on the outer particle surface, a sufficient number of silanols remain that are capable of forming siloxane bonds with the usual chloro- or alkoxysilane reagents. In this way, traditional ligands [e.g., octadecyl or octyl moieties] may be used for surface modification, thereby creating familiar, alternate selectivity. A key advantage of such particles is a significant improvement in the hydrolytic stability of their backbone at elevated pH.[116] An interesting observation is that the surface silanols in this methyl-hybrid silica are alkaline.[117]

9. The Next Plateau—UPLC Technology

The use of 3–5 μm particles was predominant in the 1990s as column packing techniques matured and column performance reproducibility improved markedly. Many of the predictions made by early pioneers were put to the test as both HPLC columns and instrumentation were refined. For example, John Knox wrote in 1977: *"The core of any high performance liquid chromatograph (HPLC) is the column. Without a column having good resolving power while operating at a high linear flow rate, all benefits from high pressure pumps, specially designed injectors, high sensitivity detectors, gradient elution devices, etc., are lost. Thus it is vital for the HPLC user to understand the main features governing column performance in the practical situation."*[118]

Knox's analysis recommended that, given the technical limitations at that time [especially detectors], the optimum column would be packed with 5-μm-diameter particles in a 5-mm-i.d. tube 10 cm in length. This column would have a reduced column length, l, of 20,000 [see Figure 9–1]. Smaller and larger particle sizes with that same L/d_p ratio would include, e.g., a 2-cm column with 1-μm particles and a 20-cm column with 10-μm particles.

Reduced Column Parameters

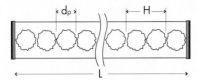

Reduced Column Length [l]:
# of particle diameters [d_p] equal to column length [L]
$$l = L/d_p$$

Reduced Plate Height [h]:
# of particle diameters [d_p] equal to height of a theoretical plate [H]
$$h = H/d_p$$

For a well-packed column , $h \approx 2$

Reduced Velocity [υ]:
$$\upsilon = u_0 d_p / D_M$$

Figure 9–1: John Knox proposed using reduced parameters [normalizing variables to particle diameter] to describe column performance and LC operational parameters so as to more easily compare columns with various dimensions and particle sizes. Note that Knox used the linear velocity u_0 based upon the elution of an unretained peak.[118] We prefer to use u_i, the linear velocity of the mobile phase in the interstitial spaces between particles.

Nearly a generation later in 1995–2003, this concept of constant reduced column length was invoked as the quest for speedier analyses led column manufacturers to introduce shorter columns [2 cm] with inner diameters ranging from 2.1–4.6 mm filled with smaller particles [1.8–5 µm]. Analyses with good separation factors normally done on longer columns now could be carried out much faster, with reduced solvent consumption [see Figure 9–2]. Ultrafast or *ballistic* gradient separations were developed using these shorter columns in tandem with mass spectrometers for biopharmaceutical assays and method development.[119]

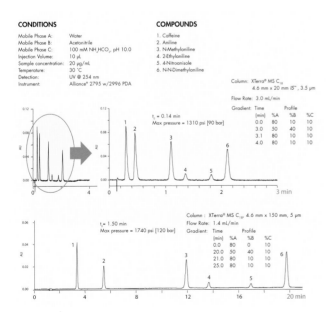

Figure 9–2: An example of how using a shorter column packed with smaller particles can be used advantageously to speed up a gradient analysis.[120]

However, fast gradients masked a multitude of issues. The extra-column contribution to band spreading became more significant with the reduction in column volume relative to that in the sample flow path [see Band Spreading in Appendix]. Since gradients tend to sharpen peak shape artificially, the loss of efficiency was most noticeable if these small columns were run under isocratic conditions. To gain the maximum speed advantage for separations, higher flow rates were needed, but as the particle size dropped, especially below 2 µm, the traditional HPLC systems suffered from back pressure limitations that prevented reaching the optimal fluid velocity within the packed bed for maximum efficiency [lowest plate height, see Chapter 10].

A recent review discusses some of these issues as well as other concerns in the context of the quest to attain the advantages of using small particles and high pressure in LC.[121] Some of the most fruitful research in this arena has been and continues to be done in the laboratories of Prof. James Jorgenson, U. North Carolina, Chapel Hill. With a particular interest in biological molecules, he and his students pioneered electrophoretic separations in capillaries in the early 1990s. That experience led him to predict that capillary-sized LC columns [i.d. << 500 µm, thereby increasing the ratio of wall area to column volume] operated at high pressure might solve the potential problem of heat dissipation due to the friction created by moving mobile phase at high velocity through small-particle packed beds.[122] He and his students set out to build a system in order to test this hypothesis.

Spherical, non-porous, octadecylsilyl-bonded silica beads, 1.5 µm in diameter, obtained from Micra Scientific were packed into 30-µm-i.d. capillary tubes 50–70 cm long and operated at pressures up to 60,000 psi [4100 bar]. At the optimum linear velocity, as many as 480,000 plates/m were obtained.[123] Jorgenson concluded that there would probably be some column diameter intermediate between the successful 30-µm capillary and a traditional 4.6–mm-i.d. bore that would also perform well at high pressure. *"Larger diameter columns would permit the use of a more universal mode of detection such as UV/visible absorbance. … It should also be possible to apply the advantages of small particles and high pressures to porous media and other modes of LC such as ion exchange and size exclusion. … note that … peaks were still several seconds wide. This suggests that ultrahigh-pressure LC could be coupled with mass spectrometry. … With the use of existing technology, … [this] joining .. may be relatively straightforward."*[123]

A year later, Jorgenson and advanced technology R&D researchers at Waters, entered into a long-term collaboration on the ultrahigh-pressure LC technique that soon became known by its acronym, UHPLC. Note, as mentioned at the end of Chapter 8, that this activity was happening independently and in parallel with the development of hybrid-particle technology packings in the Waters Chemistry R&D group.[115] Early in the new millennium, this group created a second-generation hybrid particle using a bis(triethoxysilyl)ethane monomer [see Figure 9–3].[124] The team suspected that the ethylene bridges in the backbone would serve to strengthen the new hybrid material. They found this material to be eminently suitable for LC applications. Remarkably, the resistance of its backbone to hydrolysis at high-pH was nearly an order of magnitude greater than that of the first-generation methyl hybrid.

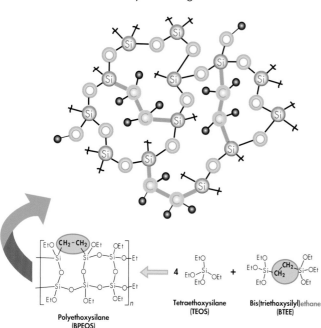

= Ethylene Bridges in Silica Matrix

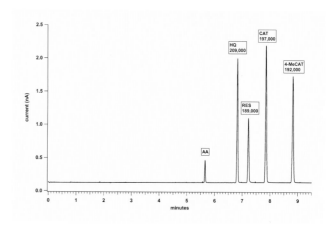

Figure 9–4a: A chromatogram run near the optimum linear velocity for a 30 µm x 49.3 cm column packed with 1.5-µm ethylene-bridged hybrid particles. Mobile phase: 50/50 CH_3CN/H_2O, 0.1% TFA. Pressure: 1600 bar [23,000 psi]. Plate counts are shown above the peaks for the retained compounds. [Reproduced with permission from reference 125.]

Tetraethoxysilane
(TEOS)

Bis(triethoxysilyl)ethane
(BTEE)

Polyethoxysilane
(BPEOS)

Figure 9–3: By combining a hexafunctional monomer with tetraethoxysilane, ethylene bridges may be distributed throughout the structure of a sturdy, stable hybrid-silica particle. This unique material's superior properties, when compared to traditional silica, proved to be the enabling technology that spurred the development of the ACQUITY UPLC® System for ultra-performance liquid chromatography.

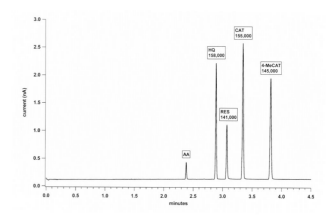

Figure 9–4b: A chromatogram run with the same column and mobile phase as in Fig. 9–4a, but at a higher flow rate. Pressure: 4400 bar [63,000 psi], the highest pressure used in these experiments. [Reproduced with permission from reference 125.]

As part of the ongoing collaboration, the Waters team produced 1.5-µm particles of this new hybrid silica and sent them to Jorgenson's team for evaluation. For six years, the latter had been using solid, non-porous silica packings as small as 1.0 µm in diameter, despite the drawback of their extremely low surface area, because no suitably strong porous packings were available. They realized, *"The problems of mass overload and low retentivity associated with nonporous silica particles could both be alleviated by the use of porous particles."*[125] Experiments [see Figure 9–4] showed the suitability of these new hybrid particles for UHPLC. *"Analysis of reduced van Deemter plots indicated that they provide similar chromatographic performance compared to nonporous silica. High-pressure runs performed on a relatively short column demonstrated that they are mechanically strong enough to withstand the high shear associated with ultrahigh pressures. ... a significant increase in retentivity was observed, necessitating the use of stronger mobile phases ... . Finally, ... ethyl-bridged hybrid particles provide the increase in sample loading capacity ... expected from their increased surface area versus nonporous particles. ... We look forward to further developments in particle synthesis that might produce more particles suitable for UHPLC."*[125]

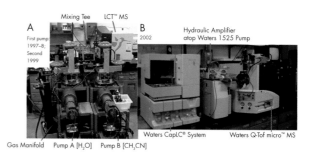

Figure 9–5: UHPLC pump engineering—from primitive to professional. A. Early separations in the Jorgenson lab were done with very large, homemade, pneumatically amplified pumps. B. As part of the collaboration with Waters, Dr. Geoff Gerhardt and former Jorgenson student Dr. Keith Fadgen greatly streamlined the pumping apparatus in 2002 by building a hydraulic amplifier to be operated in tandem with a conventional HPLC pump, boosting its pressure capability to 45,000 psi [3000 bar]. [Photographs supplied by Prof. Jorgenson, used with permission.][126]

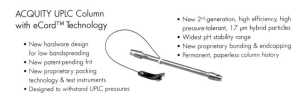

ACQUITY UPLC Column with eCord™ Technology

- New hardware design for low bandspreading
- New patent-pending frit
- New proprietary packing technology & test instruments
- Designed to withstand UPLC pressures

- New 2ⁿᵈ-generation, high efficiency, high pressure-tolerant, 1.7 μm hybrid particles
- Widest pH stability range
- New proprietary bonding & endcapping
- Permanent, paperless column history

Sample Organizer
- Accommodates vial holders & plates in any combination
- Temperature controlled environment for > 2000 samples

Solvent Reservoir Manager
- Tubing management, spill trays, & leak detection are built into fluid path throughout system

Optical Detector
- High speed stacked optical [TUV, PDA, ELSD] or adjacent MS detectors

Column Heater
- Low-dispersion solvent preheater
- Hinged for optimum placement of column outlet near detector inlet

Sample Manager
- Negligible sample carryover & reduced injection cycle times
- Pressure Assist Injection Technique

Binary Solvent Manager
- High pressure fluidics
- Minimal system volume & sample path for reduced bandspreading

Instrument Console
- Integrated system software with self-monitoring, on-board & remote diagnostics & reporting, combining intelligent device management & Connections INSIGHT™ communication via ethernet or Internet

Armed with a unique porous hybrid particle technology, the knowledge that it was possible to achieve high efficiency rapidly at high pressures, and their heritage of innovation, Waters engineers and scientists took a holistic approach to the design of a new LC system. Minimizing system volume and band spreading characteristics while dealing with the unique problems posed by moving fluids accurately, reliably, and reproducibly at high pressures—no commercial LC system had ever attempted to solve all of these problems simultaneously, blending technical sophistication and operational simplicity. The goal was not simply ultrahigh-pressure LC [UHPLC]—it was ultra performance LC [UPLC]—seeking a superior combination of speed, resolution, and sensitivity [see Figure 9–8].

Figure 9–7: Using a fine human hair as a gauge helps one to imagine the size of something as small as the 1.7-μm particles used in UPLC Columns. That, following the introduction of the first 5-μm particles, it has taken a generation to learn how to create and successfully pack such small beads into stable, reproducible, and commercially viable columns indicates the complexity of the chemistry and physics involved.

Why choose 1.7 μm and not, *e.g.*, 1.5 μm? As Jorgenson noted, the back pressure observed when pumping mobile phase through a packed bed at the optimum flow rate is inversely proportional to the *cube* of the particle diameter [$\Delta P_{opt} \propto 1/d_p^3$].[123] Therefore, the particle size was chosen as part of the holistic system design so that UPLC columns could deliver the desired performance when mated with the engineered capability to deliver mobile phase at the requisite pressures and flow rates. At the same fluid velocity, nearly 50% more pressure is generated by a 1.5-μm column than by a 1.7-μm column of equal dimensions.

Figure 9–6: Some features of the first ACQUITY UltraPerformance LC System, a product of holistic design, introduced at Pittcon 2004 in Chicago, Illinois. Sophisticated algorithms were written to augment the instrument's capability to deal with such situations as solvent compressibility, not just in pumping mobile phase, but also in sample injection. Even tube fittings had to be redesigned so as not to leak at 15,000 psi [1000 bar]. Tethered to each column is a memory chip used to store its characteristics and history.[127]

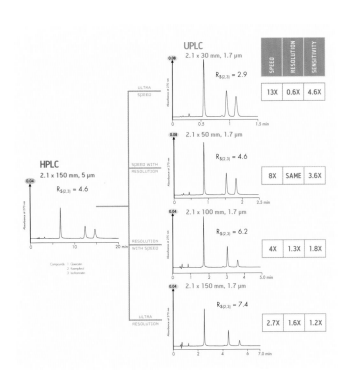

Figure 9–8: UPLC Technology affords the flexibility to optimize a separation to achieve the goal most appropriate for the desired separation. Smaller column dimensions and high efficiencies yield the narrowest peak volumes, thereby concentrating the sample bands for higher sensitivity.

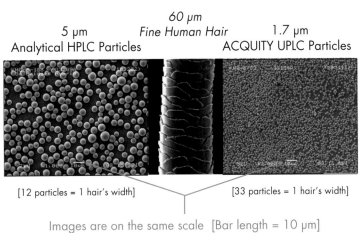

60 μm
Fine Human Hair

5 μm
Analytical HPLC Particles

1.7 μm
ACQUITY UPLC Particles

[12 particles = 1 hair's width]   [33 particles = 1 hair's width]

Images are on the same scale [Bar length = 10 μm]

As with earlier pioneering products like the M6000 pump, the initial homemade UHPLC and commercial UPLC systems embodied a new paradigm in LC performance at a time when jaded skeptics insisted that LC technology had advanced about as far as it ever was going to go. It is remarkable how strongly UPLC technology has stirred the dreams of separation scientists while simultaneously fanning the flames of competitive desires. It took very little time for those involved in the pressure cooker of drug discovery and development to seize the advantages of reduced analysis time and solvent consumption, cutting both costs and time to market while maximizing the tenure of their patent-protected revenue stream. Competitive response to UPLC technology, however, has been much slower to develop since a major commitment of multidisciplinary resources is required to build a new system holistic in scope. For those organizations that are resource-limited, after an initial flurry of predictable questions such as *'Is there a simple way to improve our current system?'* or *'How many changes must be implemented in parallel versus in series?,'* the realization sets in that fundamental LC principles both inspire and justify the perspiration required to raise even higher the bar for LC performance. Competition in the proper spirit, keeping in mind the goals of LC—achieving the desired separation as quickly as possible—leads to the accrual of benefits from a more powerful, yet already familiar, tool, that advances science and elevates the quality of life.

In the five years since the launch of ACQUITY UPLC columns and systems, dozens of improvements have been made [*e.g.*, see Figures 10–3 and 10–4] and/or are envisioned and planned. Ever mindful of the powerful effect that selectivity improvements have on the success of separations, an expanding choice of bonded phases on 1.7-µm BEH Technology™ particles, now available with two nominal average pore sizes [130 Å and 300 Å], have been created for reversed-phase and HILIC modes. Special columns and total systems have been created and qualified for the analysis of amino acids, peptides, proteins, glycans, and oligonucleotides. A second porous substrate—1.8 µm HSS or high-strength silica—has been synthesized for use at UPLC system pressures; this high-purity silica is available with three reversed-phase surface chemistries, including a novel T3 bonding and end-capping process that significantly widens the polarity range of molecules that may be retained.[128]

10. Measuring Ultra Performance— Static Plots Provide a Kinetic View

Remember that a chromatographic system is dynamic, not static. A sample is introduced into a flowing stream of mobile phase. The various analyte molecules in this sample attempt to reach both an individual and a collective equilibrium as they become distributed between the mobile and stationary phases. This equilibrium is continually reestablished as the molecules are carried by the mobile phase through the stationary phase bed to the outlet. The more times this equilibrium is established during a sample's passage through the bed, the higher is the efficiency of a column. You may consider a theoretical plate to be the zone in which this equilibrium is established. Hence there is a kinetic, as well as a thermodynamic, aspect to any measure of chromatographic performance.

Giddings used the log-log relationship between plates [N] and analysis time [t] to compare the performance of LC with GC in 1965, before the era of HPLC [see Figure 10–1]. He agreed with Knox that the use of smaller particles and higher pressures would significantly improve LC performance. Open-column LC analysis times, then, commonly were measured in hours and days. The penalty for higher efficiency was longer separation time. Note that Figure 10–1 is the first publication of what is now termed a *kinetic plot* [see Figure 10–6].

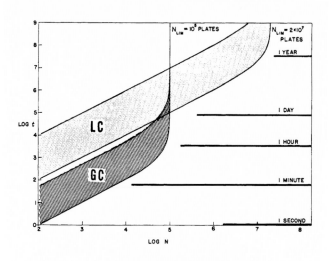

Figure 10–1: Anticipating ultra performance in liquid chromatography. Giddings entitled this figure *"Comparison of separation time in gas and liquid chromatography as a function of the number of required plates."* This figure *"should not be construed as showing the ultimate limit of chromatographic performance. With judicious changes in the systems involved, the time could undoubtedly be reduced considerably, particularly in LC. … We may tentatively conclude … where dp is subject to adjustment, the Knox hypothesis, that separation speed is increased with larger pressure drops, is correct."*[129] [Assumptions: particle size for both LC and GC: 125–149 µm; pressure: ~ 150 psi.] [Reproduced from reference 129 with permission.]

Plots of plate height versus linear velocity, the two key variables in the van Deemter equation [see Figure 5–2] are a relatively easy-to-understand tool for evaluating column performance. A curve fitted to the data points on such a plot is called a *van Deemter curve*. To compare properly the van Deemter curves for columns with different particle sizes and dimensions, it is important to use the *same* sample and separation system [mobile and stationary phases] for all measurements. Examples of such comparisons are seen in the next two figures. Figure 10–2 shows how the evolution of smaller LC particles through the decades created columns with higher efficiency. The minimum point on the van Deemter curve represents the lowest HETP or highest plate count at the corresponding optimal linear velocity. A lower minimum point and a flatter right arm on a van Deemter curve indicate higher performance.

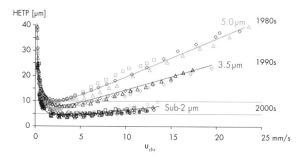

Figure 10–2: Evolution of LC Particle Technology. Actual van Deemter curves for the most common particle size used in each decade of modern LC demonstrate the progressive increase in efficiency [decrease in HETP] as well as the enhancement of separation speed at high efficiency that can be obtained with a concomitant increase in pressure capability. [Experiments by Marianna Kele.[130]]

Actual van Deemter curves, shown in Figure 10–2, illustrate how a decrease in particle size improves LC performance. Note that progress took time. Ten-μm particles, commercialized at the onset of HPLC technology in the 1970s, continued to coexist with 5-μm particles as the latter became a more viable alternative in the 1980s. The popularity of 3–3.5-μm particles increased in the 1990s since they offered an increase in performance per unit time when compared to their larger counterparts. Of course, the dramatic performance enhancement with sub-2-μm particles is the LC story of the early 21st century. This advance, as seen in Chapter 9, was only made possible in concert with a new UPLC instrument.

As indicated earlier, pioneering products are subject to continual refinement. Data presented in Figures 10–3 and 10–4 show that further technology improvements in the manufacture of sub-2-μm columns both raise performance and reduce back pressure *without* any changes to the particles themselves. These advanced columns are appropriately named *Phase II* after the improvements to the original manufacturing process.

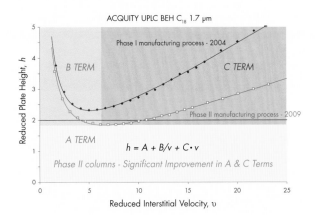

Figure 10–3: Measuring progress with the van Deemter equation. Improvements in the manufacturing process for ACQUITY UPLC BEH columns are validated by the lower mimimum point [improved A term indicates better column homogeneity] and flattened right arm [improved C term means more efficient mass transfer] of the corresponding van Deemter curve. A reduced plate height of < 2 for the Phase II columns denotes a well-packed, homogeneous column bed. Note that this van Deemter curve plots the Knox reduced plate height and velocity parameters, rather than HETP and linear velocity [see Figure 9–1.][131]

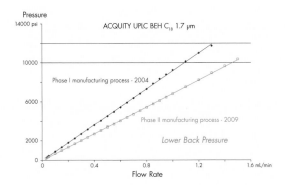

Figure 10–4: A further benefit of the new Phase II manufacturing process is lower back pressure.[131]

How this technology translates into an actual separation is shown in the plots of plate height *versus* flow rate seen in Figure 10–5. This illustration shows how most of us perceive and experience column performance. Short columns have lower plate counts, but are good for fast separations. Long columns have high plate counts, but have an operating range more limited by available pressure. Note that the optimal efficiency of a 15-cm-long, 1.7-μm Phase II ACQUITY UPLC BEH C$_{18}$ column reaches over 45,000 plates.

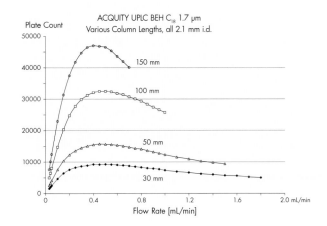

Figure 10–5: An inverse form of the van Deemter curve is obtained by plotting plate count [N], instead of plate height [HETP], *versus* volume flow rate [mL/min], instead of linear velocity [u]. The set of plots shown here demonstrate that optimal plate count is proportional to column length for Phase II ACQUITY BEH C$_{18}$ columns.[131]

Two other methods for evaluating performance are illustrated in the following series of figures. Figures 10–6, 10–7, and 10–8 are examples of kinetic plots, a tool developed by Gert Desmet and colleagues, Vrije U., Brussels.[132] Figure 10–9 shows an example of a Poppe plot, proposed by veteran separation scientist Hans Poppe, now retired from the U. Amsterdam.

Kinetic plots show the maximum column performance that may be achieved once a pressure and particle size are chosen. Figure 10–6 shows a kinetic plot [in the form of plate count *versus* analysis time]. It also shows the curves of performance [plate count] *versus* analysis time for the individual columns that make up the kinetic plot. The kinetic plot is simply the line connecting those data points where the column performance is limited by the imposed pressure constraint. We hope that, presented in this format, you may better understand the meaning of kinetic plots. Note that while they may help you to choose from an array of column choices, both axes are logarithmic. Hence kinetic plots cover a wide range of operating conditions, perhaps even beyond what you might consider practical in your situation.

Figure 10–6 attempts to explain the kinetic plot in a simple way. All the data are for a *particular* particle diameter evaluated with the *same* LC system conditions. Each curve labeled with a column length is a plot of plate count as a function of analysis time. Note that for all short columns, the plate count first increases until a maximum value is reached. Then, performance declines at even longer analysis times because longitudinal or axial diffusion processes begin to predominate [see Figures 5–2 and 5–3]. As shown previously, longer columns reach higher plate counts. However, the flow rate for each column is limited by the available pressure. Therefore, each column curve ends at a specific point on the left side of the graph. If we draw a line through the pressure-limited end points, we obtain the kinetic plot line for this particular particle size *and* these operating conditions.

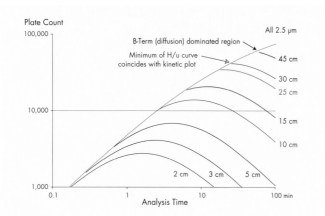

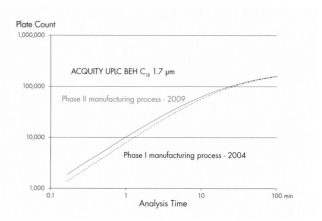

Figure 10–6: A simple explanation of a kinetic plot. The leftmost ends of the curves for each column length represent the respective points at which the maximum pressure limit has been reached when attempting to achieve the fastest analysis time. Connecting these points produces the red curve shown; this represents the kinetic plot for this particle size. Note that for the 30-cm column, the kinetic plot curve coincides with this column's maximum plate count/optimum velocity point. Above this point on the kinetic plot curve, optimum efficiency cannot be reached for longer column lengths at the now pressure-limited slower flow rates/velocities.[131,132,133]

In Figure 10–7 kinetic plots, obtained as above, are shown for three different particle sizes. Smaller particles reach a higher plate count at short analysis times while larger particles in very long columns may reach higher plate counts. However, the price paid for reaching these higher plate counts is very long analysis times. Benefits from using larger particles are only found when a single isocratic analysis takes about half a day or more. Few of us wish to devote this much time to a single analysis. Keep in mind the goals of LC: achieving the desired separation as quickly as possible.

Figure 10–8: Actual performance comparison using kinetic plot method. When the separation factor, α, is ≥ 1.05 [see Table 4–1], at equal plate counts, faster separations are achieved with the Phase II columns.[131]

While we have shown kinetic plots in a form that we believe is simple to understand, you will find most commonly in the literature plots of t/N, the *plate time, versus* the plate count, N. An example of this is shown in Figure 10–9. Interestingly, this form of the kinetic plot is the same as that drawn by Giddings [in Figure 10–1] and also used by Poppe.[134] Here, a low value of plate time on the left of the graph indicates the higher performance of the Phase II columns.

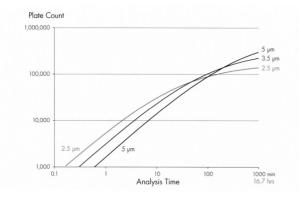

Figure 10–7: Theoretical kinetic plots for three different particle sizes. Smaller particles are best for ultra performance—achieving separations as quickly as possible.[131,132]

The performance improvement of the new Phase II manufacturing process is also evident in kinetic plots [see Figure 10–8]. The kinetic plot curve for Phase II columns is shifted upward and to the left [to higher plate counts] at fast analysis times. At long analysis times, in the diffusion-dominated region, both processes yield identical results.

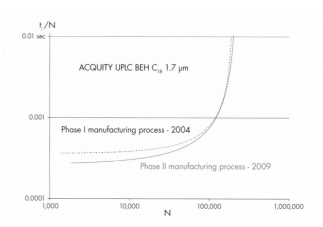

Figure 10–9: Yet another type of performance evaluation method for separation systems was proposed by veteran chromatographer Hans Poppe. He focused on the kinetic notion that the faster the speed at which a plate was generated, the more efficient would be the separation. A graph of the plate speed [t_i/N] versus the plate count [N] is now called a Poppe plot.[134] A lower curve indicates better separation performance.[131]

A detailed discussion of the relative merits of van Deemter, kinetic, and Poppe plots as tools for the measurement and optimization of separation performance is beyond the scope of our essay; you may learn more from the cited references. A van Deemter curve is perhaps the easiest for less experienced chromatographers to understand. It seems clear, however, from the evidence shown, no matter which method is used, that the smaller particle size and high-pressure capability of UPLC technology has raised the bar from high to ultra performance in LC.

11. The Essence of Success

Perhaps the truest measure of ultra performance lies not in the equations and graphs, but in the ability of scientists to apply the tools of UPLC technology with success to the solution of important problems. In five years, thousands of ACQUITY UPLC systems have been installed in laboratories all over the world. Typically within the first day of use, chromatographers have fulfilled the goals of LC with ultra performance, achieving their desired separations in dramatically less time. As Dr. Anton Jerkovich, an early UPLC adopter, said of his initial experience: *"You can achieve much better resolution, sensitivity, and speed. ACQUITY UPLC [technology] has given us the opportunity to do things that we didn't think possible with conventional HPLC. I think it has the potential to replace HPLC as we know it. To see an HPLC separation and then compare it to the UPLC [separation]—it's definitely eye-opening. The increased sensitivity has allowed us to pick up low level impurities, and that, of course, contributes to the accuracy of our method. To be able to run our analysis 5 times as fast or 8 times as fast [as] what we are currently doing with HPLC really, really helps improve our productivity. Simply running a column with small particles on a conventional HPLC is not enough. It requires a total system solution, and Waters has provided that with the ACQUITY UPLC [system]. Oh, [to] anyone that's contemplating switching from HPLC to UPLC [technology], I would just say the benefits are real, and that it really does improve your separations. I think UPLC [technology] is possibly the greatest advance in HPLC technology in decades."*[135]

Table 11–1 indicates the breadth of application areas where UPLC technology has not only replaced HPLC but in some instances accomplished analyses thought to be impossible or impractical by HPLC. So critical has UPLC technology become to chemical and pharmaceutical analysis that major companies have instituted programs to replace all HPLC instruments with UPLC systems and have standardized their analytical protocols on UPLC technology.

Hundreds of rapid, sensitive drug assay methods have been developed using UPLC technology in a multitude of pharmaceutical, biotechnology, generic drug, and contract research organization laboratories around the globe. Of course, effective sample preparation and careful creation and handling of particulate-free mobile phases, two cornerstones of good LC practice, are still important.

Chambers and her colleagues, though, have noted a key benefit of UPLC separations for these applications: *"...the added resolution of UPLC technology over HPLC yielded a statistically significant reduction in matrix effects under a variety of chromatographic conditions, and with multiple basic analytes. In summary, a combination of mixed-mode SPE, appropriate mobile phase pH, and UPLC technology results in the cleanest extracts and most sensitive and robust analytical methods for trace-level determination of drugs in plasma."*[136]

12. Quo Vadis?

Where do we go from here? Two seminal examples from disparate areas of endeavor serve to illustrate the future potential of UPLC technology. One is from a protein research laboratory. The second is happening on a factory floor.

Professor John Engen and coworkers at the Barnett Institute, Northeastern U., have perfected a dramatic analytical technique called hydrogen/deuterium exchange mass spectrometry. They use this to investigate clinically relevant protein conformation and dynamics. Keith Fadgen and Geoff Gerhardt collaborated with the Engen team to modify a Waters nanoACQUITY platform to enable UPLC separations at 0 °C.[153] Armed with this capability, and with a Waters SYNAPT™ High Definition MS system,[154] Engen and colleagues have now studied the relationship between protein dynamics and changes in protein conformation for the mutations and translocations that result in Bcr-Abl, an aberrant protein that leads to deregulated kinase activity, the oncogenic signal in chronic myelogenous leukemia [CML]. They were able to investigate the conformational changes in mutant proteins that are resistant to imatinib therapy.[155] As they state, *"Taken together, these results provide evidence that allosteric interactions and conformational changes play a major role in Abl kinase regulation in solution. Similar analyses could be performed on any protein to provide mechanistic details about conformational changes and protein function."*[156] This is groundbreaking work indeed!

Application Area	Focus
General	Reducing matrix effects in LC/MS/MS[136]
	Systematic UPLC methods development[137]
	Increasing sensitivity in bioanalytical assays[138]
Clinical analysis	Physiological amino acids[139]
	Vitamin D in serum[140]
Forensic analysis	Toxicology screening and analysis[141]
Food/Feed/Food safety analysis	Melamine & related compounds in infant formula[142]
	Amino acid analysis[143]
	Multiresidue screening/confirmation for 400+ pesticide residues in food[144]
Environmental analysis	Pharmaceuticals in water[145]
	Perfluorinated compounds in blood & plasma[146]
	Pesticides in groundwater[147]
Biopharmaceutical analysis	Characterization of monoclonal antibodies[148]
	In vitro metabolism study[149]
	Peptide analysis[150]
	Metabolic profiling[151]
	Biomarker analysis[152]

Table 11–1: A few examples of important applications of UPLC technology.

TYPICAL DOWNSTREAM BIOPHARMACEUTICAL MANUFACTURING PROCESS

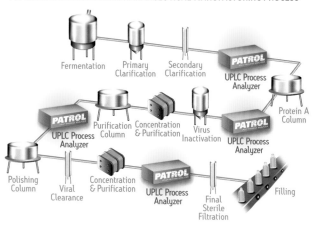

Figure 12-1: Rapid, rugged and reliable UPLC technology may now be used on factory floors by workers with minimal training to do real-time LC analysis and automatic process control. The potential impact of the resultant savings of time and materials in production is enormous.

In 2009, the first Waters PATROL™ Process Analyzer systems will be shipped to factories involved in the production of biopharmaceutical drugs. PATROL systems integrate, within a stainless-steel safety cabinet, UPLC technology with sophisticated sampling systems and control software to move QC analysis from off-line laboratories on campus to at-line or on-line locations on the production floor. The scheme is illustrated in Figure 12-1. Ultra-performance LC—achieving the desired separation as quickly as possible—is at the heart of this imminent revolution in chemical and biopharmaceutical manufacturing processes.[157]

13. Conclusion

Three Laws of Prediction:

"When a distinguished but elderly scientist states that something is possible, he is almost certainly right. When he states that something is impossible, he is very probably wrong.

The only way of discovering the limits of the possible is to venture a little way past them into the impossible.

Any sufficiently advanced technology is indistinguishable from magic."

—Arthur C. Clarke, *"Hazards of Prophecy: The Failure of Imagination,"* in *Profiles of the Future* [1973 revision]

At the dawn of the 20th century, Mikhail Tswett packed 50-μm particles into a 30-mm-long, 2-mm-i.d. transparent glass column and separated in minutes, a pair of pigments with a separation factor of 2. A century later, after packing 1.7-μm particles under high pressure into an opaque steel column with the same dimensions, a far more difficult separation, with $\alpha = 1.1$, may be achieved in mere seconds using at least one million times less sample and sophisticated detection technologies for quantitative analysis and structure elucidation. To the uninitiated, both events seem magical. Each involved pushing past the possible into the realm of the seemingly impossible. And anyone who thought them impossible was most certainly wrong.

In this decade, a renewed sense of urgency—to combat disease; to preserve our environment; to protect our food supply; to guard our drinking water; to secure our world; to prolong peace—drives the need for greater efficiency, sensitivity, speed, sophistication, and confidence in our analytical arsenal. Reliable, reproducible UPLC technology fills that need. It has been made by the best people with the simplest, yet most sophisticated, designs that those engineers and chemists can devise; it is affordable to the many laboratories that need it; and it will enable science to generate an abundance of blessings for our global human family.

Don't ever think for a moment, however, that we have seen the end of innovation and revolution. As awe-inspiring as the initial efforts have been in creating the first ultra-performance LC Systems, it will be even more wonderful to watch the refinement of this technology in the next decade, to be followed inevitably by another iteration in the cycle of revolution and evolution. Now, at the dawn of the 21st century, a suitable foundation has been secured for the advancement of separation science. Let us resolve to do all the things of which we are capable so that, as a great inventor once suggested, we can *"literally astound ourselves."*[158]

14. Resources

To learn more about UPLC technology, please take advantage of the variety of resources available on the web, most at waters.com:

- Read *A Beginner's Guide to UPLC: Ultra-Performance Liquid Chromatography*: by Eric S. Grumbach, Joseph C. Arsenault, and Douglas R. McCabe; for ordering information, search for **715002099** on waters.com .

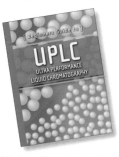

- Review case studies of the benefits, both scientific and financial, from successful implementation of UPLC technology. Follow the link at waters.com/uplc .

- Join the Waters ACQUITY UPLC Online Community, a place to learn, share, and advance your chromatography skills. Registration is free at: http://www.waters.com/myuplc .

- Review a bibliography of > 500 publications in peer-reviewed journals that cite the use of UPLC technology [PDF available, search for **720001368en** on waters.com .

- Find > 3600 references on GOOGLE® Scholar http://scholar.google.com . Use this search string exactly as written: **UPLC OR UHPLC**. Add another term to this string to seek specific applications: *e.g.*, a search for **(UPLC OR UHPLC) aflatoxin** returns about 50 references. HINT: Bookmark the URL for your GOOGLE Scholar search and check it again on a regular basis to turn up newly added citations. The growth rate of citations since 2004 continues to be exponential.

- For the most current product information and related material, visit both waters.com/acquity and waters.com/acquitycolumns .

Appendix: LC Nomenclature

*Indicates a definition adapted from: L.S. Ettre, *Nomenclature for Chromatography, Pure Appl. Chem.* **65**: 819-872 [1993], © 1993 IUPAC; an updated version of this comprehensive report is available in the *Orange Book*, Chapter 9: *Separations* [1997] at: http://www.iupac.org/publications/analytical_compendium .

Alumina

A porous, particulate form of aluminum oxide [Al_2O_3] used as a stationary phase in normal-phase adsorption chromatography. Alumina has a highly active basic surface; the pH of a 10% aqueous slurry is about 10. It is successively washed with strong acid to make neutral and acidic grades [slurry pH 7.5 and 4, resp.]. Alumina is more hygroscopic than silica. Its activity is measured according to the Brockmann[11] scale for water content; *e.g.,* Activity Grade I contains 1% H_2O.

Band [see Peak]

Band Broadening [see Band Spreading]

Band Spreading

Sometimes called band broadening, band dispersion, peak broadening, or peak dispersion. This is the process whereby the concentration profile of a sample component, once formed upon injection into a chromatographic system, spreads due to: (1) diffusion processes [eddy diffusion, axial diffusion, mass transfer] within the packed bed [column band spreading, see Figures 5–1 and 5–2]; (2) collective *hydrodynamic* effects outside of the packed bed on the flow path between the injector and detector cell [extra-column band spreading, see Figure A–1]; or (3) peak distortion from electronic sources such as an improper detector time constant and/or inadequate acquisition rates in the data collection system. Holistic system design aims to minimize or eliminate all three sources of band spreading, particularly the latter two [see Figure 9–6].[159]

Bar

Bar is the unit of pressure recognized in the European Union. It is not an SI [International System of Units, metric system] or a cgs [centimeter-gram-second] unit, but is accepted for metric use. It is nearly equal to one atmosphere: 1 bar = 0.98692 atm. This unit was introduced by British meteorologist Sir Napier Shaw in 1909 and internationally adopted in 1929. In England and the United States, the common unit of pressure is psi [pounds per square inch]: 1 bar = 14.5037744 psi. A bar is related to another unit of pressure, the Pascal: 10 bar = 1 megaPascal [MPa]. Throughout this volume we will write pressure in both approximate bar and psi units.

Baseline*

The portion of the chromatogram recording the detector response when only the mobile phase emerges from the column.

Calcination [v. calcine]

A process whereby a material is heated to a high temperature, without fusing or melting it, in order to drive off volatile matter or to effect chemical changes such as oxidation or dehydration. Organic impurities bound to the surface of active solids such as magnesia or silica may be burned completely to carbon dioxide and water to produce residue-free sorbents for pesticide analysis or other high-sensitivity, trace LC methods. If insufficient oxygen is present, organic matter may be carbonized, leaving behind a carbon layer on the substrate.

Calcination is also carried out just below the fusion point of silica to form covalent siloxane bonds between adjacent silanols with concomitant loss of water:

$$-Si-OH \ + \ HO-Si- \ \xrightarrow[-H_2O]{\geq 900\ °C} \ -Si-O-Si-$$

This dehydration process may be used to coalesce micropores and reduce surface area, or to bond adjacent colloidal silica particles at their contact points, thereby hardening or strengthening the overall silica particle.

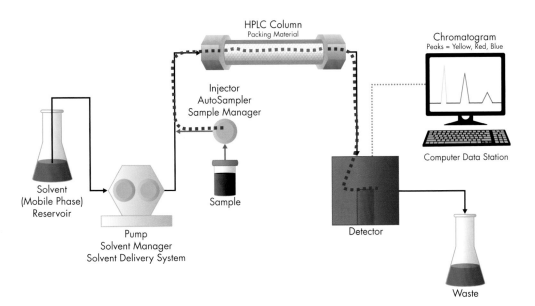

Figure A–1: Band spreading may occur anywhere along the sample path from point of injection to point of detection in an LC system. *Column* band spreading occurs within the packed bed [red dotted line]. *Extra-column* band spreading occurs in the injector, column end fittings, detector cell, and all fluid transport tubing and fittings connecting each element [blue dotted line] as well as in the process of data collection.

Cartridge

A type of column, without endfittings, that consists simply of an open tube wherein the packing material is retained by a frit at either end. SPE cartridges may be operated in parallel on a vacuum-manifold. HPLC cartridges are placed into a cartridge holder that has fluid connections built into each end. Cartridge columns are easy to change, less expensive, and more convenient than conventional columns with integral endfittings.

Chromatogram*

A graphical or other presentation of detector response or other quantity used as a measure of the concentration of the analyte in the effluent versus effluent volume or time. In planar chromatography [e.g., thin-layer chromatography or paper chromatography], *chromatogram* may refer to the paper or layer containing the separated zones.

Chromatography*

A dynamic physicochemical method of separation in which the components to be separated are distributed between two phases, one of which is stationary [*the stationary phase*] while the other [the *mobile phase*] moves relative to the stationary phase.

Coacervation [verb: coacervate]

A process whereby a suspension of colloidal particles is admixed into another partially miscible or immiscible medium, forming droplets. If the particles are composed of organic monomers, then, while in suspension, the polymerization process may be catalyzed to form spherical polymer beads. If the medium is dessicating and the particles are silica, such a sol-gel process may be used to create spherical silica beads. Subsequent drying and calcination steps yield porous silica particles of a size and with characteristics suitable for LC separations. An example of the product of a novel process involving coacervation of a mixture of organic monomers and tiny silica particles is described in Figure 8–2.

Column Volume* [Vc]

The geometric volume of the part of the tube that contains the packing [internal cross-sectional area of the tube multiplied by the packed bed length, L]. The *interparticle volume* of the column, also called the *interstitial volume*, is the volume occupied by the mobile phase between the particles in the packed bed. The void volume [V_0] is the total volume occupied by the mobile phase, *i.e.* the sum of the interstitial volume and the *intraparticle volume* [also called *pore volume*].

Detector* [see Sensitivity]

A device that indicates a change in the composition of the eluent by measuring physical or chemical properties [e.g., UV/visible light absorbance, differential refractive index, fluorescence, or conductivity]. If the detector's response is linear with respect to sample concentration, then, by suitable calibration with standards, the amount of a component may be quantitated. Often, it may be beneficial to use two different types of detectors in series. In this way, more corroboratory or specific information may be obtained about the sample analytes. Some detectors [e.g., electrochemical, mass spectrometric] are destructive; *i.e.*, they effect a chemical change in the sample components. If a detector of this type is paired with a non-destructive detector, it is usually placed second in the flow path.

Diffusion*

Molecular diffusion is the net transport of molecules that results from their molecular motions alone in the absence of turbulent mixing; it occurs when there is a concentration gradient. Components tend to move from regions of high concentration to low concentration so as to seek equilibrium. Larger molecules [e.g., proteins, polymers] diffuse much more slowly than small molecules. The laws governing the transfer of mass by diffusion were first proposed by physiologist Adolf Fick in 1855.[160]

Diffusion Coefficient [D]*

The diffusion coefficient (D) is the amount of a particular substance that diffuses across a unit area in 1 second under the influence of a gradient of one unit. It is usually expressed in the units cm^2s^{-1}. D_S is the diffusion coefficient characterizing the diffusion in the stationary phase. D_M is the diffusion coefficient characterizing the diffusion in the mobile phase. Diffusion coefficients for large molecules are significantly smaller [up to 100 × less] than those for low molecular-weight molecules. Hence, optimal UPLC separations of biomolecules like proteins are performed on longer columns at lower flow rates.

Discrimination Factor [d_0, see Resolution]

Two common scenarios limit the utility of resolution as a measure of separation quality: (1) R_s may increase indefinitely as the distance between peaks widens; and (2) R_s may be greater than one for a pair of peaks, differing widely in peak height and area, yet showing no visible separation. A superior criterion for peak separation is the discrimination factor, introduced by Zoubair El Fallah and Michel Martin, shown in Figure A–2.[161]

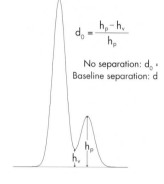

Discrimination Factor, d_0

$$d_0 = \frac{h_p - h_v}{h_p}$$

No separation: $d_0 =$
Baseline separation: d

Figure A–2: The Discrimination Factor is measured by using the height of the shortest peak [h_p] and the height above baseline at the valley between the adjacent peaks [h_v]. The closer d_0 is to unity, the more complete is the separation between peaks. As Neue points out, d_0 has significant advantages: ease of calculation for computerized data systems; good sensitivity to both peak shape and relative peak heights; and as a rational goal for optimal method development.[162]

Dispersion [see Band Spreading]

Display

A device that records the electrical response of a detector on a computer screen in the form of a chromatogram. Advanced data recording systems also perform calculations using sophisticated algorithms, e.g., to integrate peak areas, subtract baselines, match spectra, quantitate components, and identify unknowns by comparison to standard libraries.

Eddy Diffusion*

An eddy is a portion of fluid that moves in some way differently from the main flow. Eddy diffusion or eddy dispersion is a process by which substances are mixed in a fluid system due to eddy motion.

Efficiency [H, see Plate Number, Resolution, Sensitivity, Speed]

A measure of a column's ability to resist the dispersion of a sample band as it passes through the packed bed. An efficient column minimizes *band dispersion* or *band spreading*. Higher efficiency is important for effective separation, greater sensitivity, and/or identification of similar components in a complex sample mixture.

Nobelists Martin and Synge, by analogy to distillation, introduced the concept of *plate height* [H, or H.E.T.P., *height equivalent to a theoretical plate*] as a measure of chromatographic efficiency and as a means to compare column performance.[22] Presaging HPLC and UPLC technology, they recognized that a homogeneous bed packed with the smallest possible particle size [requiring higher pressure] was key to maximum efficiency. The relation between column and separation system parameters that affect band spreading was later described in an equation by Van Deemter [see Figure 5–2].[36]

Chromatographers often refer to a quantity that they can calculate easily and directly from measurements made on a chromatogram, namely *plate number* [N], as efficiency. Plate height is then determined from the ratio of the length of the column bed to N [H = L/N; methods of calculating N from a chromatogram are shown in Figure 5–1]. It is important to note that calculation of N or H using these methods is correct only for isocratic conditions and cannot be used for gradient separations.

Eluate

The portion of the *eluent* that emerges from the column outlet containing analytes in solution. In analytical HPLC, the *eluate* is examined by the detector for the concentration or mass of analytes therein. In preparative HPLC, the eluate is collected continuously in aliquots at uniform time or volume intervals, or discontinuously only when a detector indicates the presence of a peak of interest. These fractions are subsequently processed to obtain purified compounds.

Eluent

The *mobile phase* [see Elution Chromatography].

Eluotropic Series

A list of solvents ordered by *elution strength* with reference to specified analytes on a standard sorbent. Such a series is useful when developing both isocratic and gradient elution methods. Trappe coined this term after showing that a sequence of solvents of increasing polarity could separate lipid fractions on alumina.[163] Later, Snyder measured and tabulated solvent strength parameters for a large list of solvents on several normal-phase LC sorbents.[164] Neher created a very useful *nomogram* by which *equi-eluotropic* [constant elution strength] mixtures of normal-phase solvents could be chosen to optimize the selectivity of TLC separations.[165]

A typical normal-phase *eluotropic series* would start at the weak end with non-polar aliphatic hydrocarbons, *e.g.,* pentane or hexane, then progress successively to benzene [an aromatic hydrocarbon], dichloromethane [a chlorinated hydrocarbon], diethyl ether, ethyl acetate [an ester], acetone [a ketone], and, finally, methanol [an alcohol] at the strong end.

Elute* [verb]

To chromatograph by elution chromatography. The process of elution may be stopped while all the sample components are still on the chromatographic bed [planar thin-layer or paper chromatography] or continued until the components have left the chromatographic bed [column chromatography].

Note: The term *elute* is preferred to *develop* [a term used in planar chromatography], to avoid confusion with the practice of method development, whereby a separation system [the combination of mobile and stationary phases] is optimized for a particular separation.

Elution Chromatography*

A procedure for chromatographic separation in which the mobile phase is continuously passed through the chromatographic bed. In HPLC, once the detector baseline has stabilized and the separation system has reached equilibrium, a finite slug of sample is introduced into the flowing mobile phase stream. Elution continues until all analytes of interest have passed through the detector.

Elution Strength

A measure of the affinity of a solvent relative to that of the analyte for the stationary phase. A weak solvent cannot displace the analyte, causing it to be strongly retained on the stationary phase. A strong solvent may totally displace all the analyte molecules and carry them through the column unretained. To achieve a proper balance of effective separation and reasonable elution volume, solvents are often blended to set up an appropriate *competition* between the phases, thereby optimizing both selectivity and separation time for a given set of analytes [see Selectivity].

Dipole moment, dielectric constant, hydrogen bonding, molecular size and shape, and surface tension may give some indication of elution strength. Elution strength is also determined by the separation mode. An *eluotropic series* of solvents may be ordered by increasing strength in one direction under *adsorption* or *normal-phase* conditions; that order may be nearly opposite under *reversed-phase partition* conditions.

Fluorescence Detector

Fluorescence detectors *excite* a sample with a specified wavelength of light. This causes certain compounds to fluoresce and emit light at a higher wavelength. A sensor, set to a specific *emission wavelength* and masked so as not to be blinded by the excitation source, collects only the emitted light. Often analytes that do not natively fluoresce may be derivatized to take advantage of the high sensitivity and selectivity of this form of detection, *e.g.,* AccQ•Tag™ derivatization of amino acids.

Flow Rate* [see Linear Velocity]

The volume of mobile phase passing through the column in unit time. In HPLC systems, the flow rate is set by the controller for the solvent delivery system [pump]. Flow rate accuracy can be checked by timed collection and measurement of the effluent at the column outlet. Since a solvent's density varies with temperature, any calibration or flow rate measurement must take this variable into account. Most accurate determinations are made, when possible, by weight, not volume.

Uniformity [precision] and *reproducibility* of flow rate is important to many LC techniques, especially in separations where *retention times* are key to analyte identification, or in *gel-permeation chromatography* where calibration and correlation of retention times are critical to accurate molecular-weight-distribution measurements of polymers.

Often, separation conditions are compared by means of *linear velocity*, not flow rate. The linear velocity is calculated by dividing the flow rate by the cross-sectional area of the column [or a fraction thereof]. While flow rate is expressed in volume/time [*e.g.,* mL/min], linear velocity is measured in length/time [*e.g.,* mm/sec].

Gel-Permeation Chromatography*

Separation based mainly upon exclusion effects due to differences in molecular size and/or shape. *Gel-permeation chromatography* and *gel-filtration chromatography* describe the process when the stationary phase is a swollen gel. Both are forms of *size-exclusion chromatography*. Porath and Flodin first described gel filtration using dextran gels and aqueous mobile phases for the size-based separation of biomolecules.[51] Moore applied similar principles to the separation of organic polymers by size in solution using organic-solvent mobile phases on porous polystyrene-divinylbenzene polymer gels.[50]

Gradient

The change over time in the relative concentrations of two [or more] miscible solvent components that form a mobile phase of increasing elution strength. A *step gradient* is typically used in solid-phase extraction; in each step, the eluent composition is changed abruptly from a weaker mobile phase to a stronger mobile phase. It is even possible, by drying the SPE sorbent bed in between steps, to change from one solvent to another immiscible solvent.

A *continuous* gradient is typically generated by a low- or high-pressure mixing system according to a pre-determined curve [linear or non-linear] representing the increase in concentration of the stronger solvent B in the initial solvent A over a fixed time period. A *hold* at a fixed *isocratic* solvent composition can be programmed at any time point within a continuous gradient. At the end of a separation, the *gradient program* can also be set to return to the initial mobile phase composition to re-equilibrate the column in preparation for the injection of the next sample. Sophisticated HPLC systems can blend as many as four or more solvents [or solvent mixtures] into a continuous gradient [see Figure A–3].

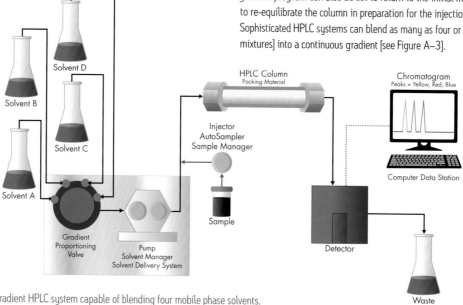

Figure A–3: Low-pressure-gradient HPLC system capable of blending four mobile phase solvents.

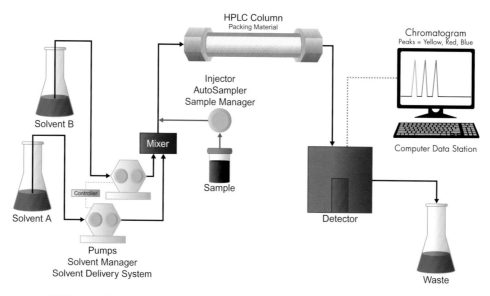

Figure A–4: High-pressure-gradient HPLC system with two-solvent mixing capability.

Injector [Autosampler, Sample Manager]

A mechanism for accurately and precisely introducing [*injecting*] a discrete, predetermined volume of a sample solution into the flowing mobile phase stream. The injector can be a simple manual device, or a sophisticated autosampler that can be programmed for unattended injections of many samples from an array of individual vials or wells in a predetermined sequence. Sample compartments in these systems may even be temperature controlled to maintain sample integrity over many hours of operation.

Most modern injectors incorporate some form of syringe-filled sample loop that can be switched on- or offline by means of a multi-port valve. A well-designed, minimal-internal-volume injection system is situated as close to the column inlet as possible and minimizes the spreading of the sample band. Between sample injections, it is also capable of being flushed to waste by mobile phase, or a wash solvent, to prevent *carryover* [contamination of the present sample by a previous one].

Samples are best prepared for injection, if possible, by dissolving them in the mobile phase into which they will be injected; this may prevent issues with separation and/or detection. If another solvent must be used, it is desirable that its elution strength be equal to or less than that of the mobile phase. It is often wise to mix a bit of a sample solution with the mobile phase offline to test for precipitation or miscibility issues that might compromise a successful separation.

Inlet

The end of the column bed where the mobile phase stream and sample enter. A porous, inert frit retains the packing material and protects the sorbent bed inlet from particulate contamination. Good LC practice dictates that samples and mobile phases should be particulate-free; this becomes imperative for small-particle columns whose inlets are much more easily plugged. If the column bed inlet becomes clogged and exhibits higher-than-normal *back pressure*, sometimes, reversing the flow direction while directing the effluent to waste may dislodge and flush out sample debris that sits atop the frit. If the debris has penetrated the frit and is lodged in the inlet end of the bed itself, then the column has most likely reached the end of its useful life. Beware of aqueous mobile phases in which bacteria colonies will grow.

Ion-Exchange Chromatography[*]

This separation mode is based mainly on differences in the ion-exchange affinities of the sample components. Separation of primarily inorganic ionic species in water or buffered aqueous mobile phases on small particle, superficially porous, high-efficiency, ion-exchange columns followed by conductometric or electrochemical detection is referred to as ion chromatography [IC].

Isocratic Elution[*]

A procedure in which the composition of the mobile phase remains constant during the elution process. A typical isocratic system schematic is shown in Figure A–5.

Linear Velocity [see Flow Rate]

Nominally the mobile phase volume flow rate divided by the cross-sectional area of a column bed [or a fraction thereof]. Linear velocity is expressed in units of distance/time. There is some confusion about which linear velocity should be used in hydrodynamic calculations.[166] We use the interstitial velocity, or the rate at which mobile phase flows through the spaces around the particles in a packed bed [see Figure 9–1].

Liquid Chromatography[*] [LC]

A separation technique in which the mobile phase is a liquid. Liquid chromatography can be carried out either in a column or on a plane [TLC or paper chromatography]. Modern liquid chromatography utilizing smaller particles and higher inlet pressure was termed *high-performance* (or *high-pressure*) *liquid chromatography* [HPLC] in 1970. In 2004, *ultra-performance liquid chromatography* dramatically raised the performance of LC to a new plateau [see UPLC Technology].

Mass Transfer [see Diffusion][*]

A spontaneous process of transfer of molecules by diffusion within or between phases such as liquids and solids. The driving force, especially in liquids, can be a difference in concentration; systems naturally tend to seek equilibrium.

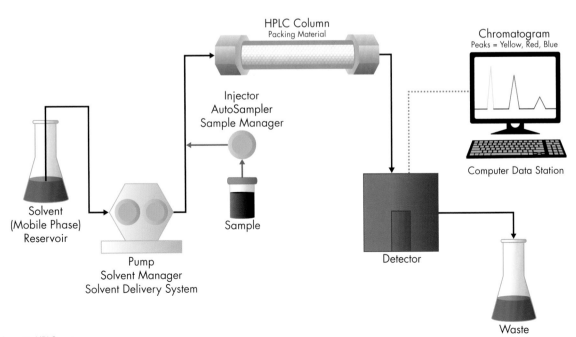

Figure A–5: Isocratic HPLC system.

Mobile Phase* [see Eluate, Eluent]

A fluid that percolates, in a definite direction, through the length of the stationary-phase sorbent bed. The mobile phase may be a liquid [*liquid chromatography*] or a gas [*gas chromatography*] or a supercritical fluid [*supercritical-fluid chromatography*]. In gas chromatography the expression *carrier gas* may be used for the mobile phase. In elution chromatography, the mobile phase may also be called the *eluent*, while the word *eluate* is defined as the portion of the mobile phase that has passed through the sorbent bed and contains the compounds of interest in solution.

Morphology

Chemists have borrowed this term from the area of biology that deals with the form and structure of plants and animals. We use it as an inclusive term that includes the overall shape, as well as the distribution and magnitude of particle properties such as pore volume, pore dimensions, pore shape, and surface area.

Normal-Phase Chromatography*

An elution procedure in which the stationary phase is more polar than the mobile phase. This term is used in liquid chromatography to emphasize the contrast to *reversed-phase chromatography*.

Peak* [see Plate Number]

The portion of a chromatogram recording the detector response while a single component is eluted from the column. If separation is incomplete, two or more components may be eluted as one *unresolved* peak. Peaks eluted under optimal conditions from a well-packed, efficient column, operated in a system that minimizes band spreading, approach the shape of a Gaussian distribution. Quantitation is usually done by measuring the *peak area* [enclosed by the baseline and the peak curve]. Less often, peak height [the distance measured from the peak apex to the baseline] may be used for quantitation. This procedure requires that both the peak width and the peak shape remain constant.

Plate Number* [N, see Efficiency]

A number indicative of column performance [mechanical separation power or efficiency, also called *plate count, number of theoretical plates,* or *theoretical plate number*]. It relates the magnitude of a peak's retention to its width [*variance* or *band spread*]. In order to calculate a plate count, it is assumed that a peak can be represented by a Gaussian distribution [a statistical *bell curve*]. At the inflection points [60.7% of peak height], the width of a Gaussian curve is twice the *standard deviation* [sigma, σ] about its mean [located at the peak apex]. As shown in Figure 5–1, a Gaussian curve's peak width measured at other fractions of peak height can be expressed in precisely defined multiples of σ. Peak retention [retention volume, V_R, or retention time, t_R] and peak width must be expressed in the same units, because N is a dimensionless number. Note that the five-sigma method of calculating N is a more stringent measure of column homogeneity and performance, as it is more severely affected by peak asymmetry.[167] Computer data stations can automatically delineate each resolved peak and calculate its corresponding plate number.

Preparative Chromatography

The process of using liquid chromatography to isolate a compound in a quantity and at a purity level sufficient for further experiments or uses. For pharmaceutical or biotechnological purification processes, columns several feet in diameter can be used for multiple kilograms of material. For isolating just a few micrograms of a valuable natural product, an analytical HPLC column is sufficient [see Figure A–6]. Both are preparative chromatographic approaches, differing only in scale.

Recycle

The recycle technique is a way to gain the benefits of a longer column [more plates, more capacity] without the drawback of increasing back pressure. In this process, a portion of the column eluate containing an incompletely separated mixture of sample components is diverted from the waste stream and passed back through the pump and column repetitively until the desired resolution is attained. This process is useful for preparative LC [see the example in Figure 6–7] and was promoted in the early 1970s in several seminal application notes written by Jim Waters. A variation of this technique, known as *alternate-column recycle*, uses a pair of columns, a six-port

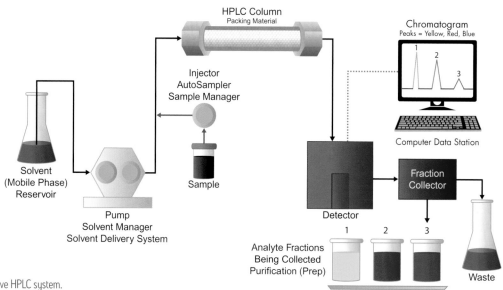

Figure A–6: Preparative HPLC system.

switching valve, and, optionally, a pair of pressure-resistant detector cells. This mode was developed in 1971 for GPC separations. It has the advantage of not sending the sample back through the pump during each cycle. Instead, the sample is switched alternately from one column to the other.[168] In 1982, this technique was first used for analytical GPC separations of small molecules, achieving 400,000 plates in about three hours.[169] Independently, 25 years later, it has been investigated anew, renamed, and applied to UHPLC.[170] Using alternate-column recycle, it is possible to generate a very high plate count; the price for this high efficiency, of course, is the time required for multiple column passes. Also, alternate-column recycle continually uses solvent [the classical recycle technique conserves mobile phase during the time the eluate is being recycled].

Resolution* [R_s, see Discrimination Factor, Selectivity]

The separation of two adjacent peaks, expressed as the difference in their corresponding retention times, divided by their average peak width at the baseline. R_s = 1.0 indicates that the tangents drawn to the inflection points on adjacent peaks cross at the baseline [See Figure A–7]. R_s = 1.25 indicates that two peaks of equal width are nearly separated at the baseline. When R_s = 0.6, the only visual indication of the presence of two peaks on a chromatogram is a small notch near the peak apex. Resolution becomes a less useful measure of separation when adjacent peaks differ greatly in height. A better method than *resolution* for measuring the quality of a separation is to use the *discrimination factor* [see Figure A–2].

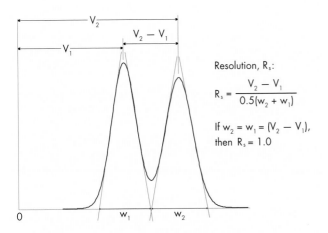

Resolution, R_s:

$$R_s = \frac{V_2 - V_1}{0.5(w_2 + w_1)}$$

If $w_2 = w_1 = (V_2 - V_1)$, then $R_s = 1.0$

Substituting the equation for N [using retention volume, not time], in Figure 5–1 into the expression above for R_s, the following relationship is obtained:

$$R_s = N^{0.5} \cdot \frac{k_1}{(k_1 + 1)} (\alpha - 1)$$

Figure A–7: The simplified resolution equation shown here indicates the relative importance of key separation parameters. Higher efficiency columns produce narrower peaks and improve resolution for difficult separations; however, *resolution* increases by only the *square root of N*. The most powerful method of improving a separation, thereby increasing resolution, is to increase *selectivity* by altering the mobile/stationary phase combination used for the chromatographic separation [see Table 4–1].

Retention Factor* [k]

A measure of the time the sample component resides in the stationary phase relative to the time it resides in the mobile phase; it expresses how much longer a sample component is retarded by the stationary phase than it would take to travel through the column with the velocity of the mobile phase. Mathematically, it is the ratio of the adjusted retention time [volume] and the hold-up time [volume]: $k = t_R'/t_M$ [see Retention Time and Selectivity].

Note: In the past, this term has also been expressed as *partition ratio*, *capacity ratio*, *capacity factor*, or *mass distribution ratio* and symbolized by k'.

Retention Time* [t_R]

The time between the start of elution [typically, in HPLC, the moment of injection or sample introduction] and the emergence of the peak maximum. The *adjusted retention time*, t_R', is calculated by subtracting from t_R the *hold-up time* [t_M, the time from injection to the elution of the peak maximum of a totally unretained analyte].

Reversed-Phase Chromatography*

An elution procedure used in liquid chromatography in which the mobile phase is significantly more polar than the stationary phase, *e.g.* a microporous silica-based material with alkyl chains chemically bonded to its accessible surface. Note: Avoid the incorrect term *reverse phase*.[171]

Selectivity [Separation Factor, α]

A term used to describe the magnitude of the difference between the relative thermodynamic affinities of a pair of analytes for the specified mobile and stationary phases that comprise the separation system. The proper term is *separation factor* [α]. It equals the ratio of retention factors, k_2/k_1 [see Retention Factor]; by definition, α is always ≥ 1. If α = 1, then both peaks co-elute, and no separation is obtained, no matter how many plates are available in a column. It is important in preparative chromatography to maximize α for highest sample loadability and throughput.

Sensitivity* [S]

The signal output per unit concentration or unit mass of a substance in the mobile phase entering the detector, *e.g.*, the slope of a linear calibration curve [see Detector]. For concentration-sensitive detectors [*e.g.*, UV/VIS absorbance], sensitivity is the ratio of peak height to analyte concentration in the peak. For mass-flow-sensitive detectors, it is the ratio of peak height to unit mass. If sensitivity is to be a unique performance characteristic, it must depend only on the chemical measurement process, not upon scale factors.

The ability to detect [*qualify*] or measure [*quantify*] an analyte is governed by many instrumental and chemical factors. Well-resolved peaks [maximum selectivity] eluting from high efficiency columns [narrow peak width with good symmetry for maximum peak height] as well as good detector sensitivity and specificity are ideal. Both the separation system interference and electronic component noise should also be minimized to achieve maximum sensitivity.

Solid-Phase Extraction [SPE]

A sample preparation technique that uses LC principles to isolate, enrich, and/or purify analytes from a complex matrix applied to a miniature chromatographic bed. *Offline* SPE is done [manually or *via* automation] with larger particles in individual plastic cartridges or in micro-elution plate wells, using low positive pressure or vacuum to assist flow [see Figure A–8]. *Online* SPE is done with smaller particles in miniature HPLC columns using higher pressures and a valve to switch the SPE column online with the primary HPLC column, or offline to waste, as appropriate.

SPE methods use step gradients [see Gradient] to accomplish bed conditioning, sample loading, washing, and elution steps. Samples are loaded typically under conditions where the k of important analytes is as high as possible, so that they are fully retained during loading and washing steps. Elution is then done by switching to a much stronger solvent mixture [see Elution Strength]. The goal is to remove matrix interferences and to isolate the analyte in a solution, and at a concentration, suitable for subsequent analysis.

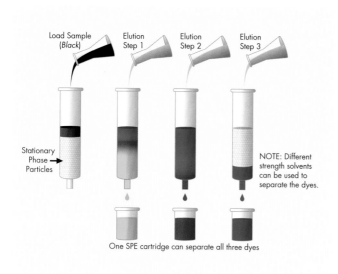

Figure A–8: An example of solid-phase extraction [SPE] using a step gradient to isolate different species from a sample mixture.

Speed [see Efficiency, Flow Rate, Linear Velocity, Resolution]

A benefit of operating LC separations at higher *linear velocities* using smaller-volume, smaller-particle analytical columns, or larger-volume, larger-particle preparative columns. Order-of-magnitude advances in LC speed came in 1972 [with the use of 10 μm particles and pumps capable of delivering accurate mobile-phase flow at 6000 psi (400 bar)], in 1976 [with 75-μm preparative columns operated at a flow rate of 500 mL/min], and in 2004 [with the introduction of UPLC technology—1.7-μm-particle columns operated at 15,000 psi (1000 bar)].[49]

High-speed analytical LC systems must not only accommodate higher pressures throughout the fluidics; injector cycle time must be short; gradient mixers must be capable of rapid turnaround between samples; detector sensors must rapidly respond to tiny changes in eluate composition; and data systems must collect the dozens of points each second required to plot and to quantitate narrow peaks accurately.

Together, higher resolution, higher speed, and higher efficiency typically deliver higher *throughput*. More samples can be analyzed in a workday. Larger quantities of compound can be purified per run or per process period.

Stationary Phase*

One of the two phases forming a chromatographic system. It may be a solid, a gel, or a liquid. If a liquid, it may be distributed on a solid. This solid may or may not contribute to the separation process. The liquid may also be chemically bonded to the solid [*bonded phase*, see Figure 8–1] or immobilized onto it [*immobilized phase*].

The expression *chromatographic bed* or *sorbent* may be used as a general term to denote any of the different forms in which the stationary phase is used.

The use of the term *liquid phase* to denote the *mobile phase* in LC is discouraged. This avoids confusion with gas chromatography where the *stationary phase* is called a *liquid phase* [most often a liquid coated on a solid support].

Open-column liquid-liquid partition chromatography [LLPC] did not translate well to HPLC. It was supplanted by the use of bonded-phase packings. LLPC proved incompatible with modern detectors because of problems with bleed of the stationary-phase-liquid coating off its solid support, thereby contaminating the immiscible liquid mobile phase.

UPLC Technology

The use of a high-efficiency LC system holistically designed to accommodate sub-2 μm particles and very high operating pressure is termed *ultra-performance liquid chromatography* [UPLC technology].[172] The major benefits of this technology are significant improvements in resolution and/or sensitivity over HPLC, and/or faster run times while maintaining the resolution seen in an existing HPLC separation.

Xerogel

A term used for the dried out open solid structures which have passed a *gel* stage during preparation (*e.g.*, silica gel); and also for dried out compact macromolecular gels such as gelatin or rubber. During drying, shrinkage is unhindered, and the solid is characterized by its small pores, high surface area, and high total porosity.

Epilogue: Connections—All in the Family

This volume puts some highlights of LC history in perspective, paying particular attention to column technology, while seeking the origins of ultra performance. An essay of this size perforce may not be a comprehensive treatise. Many marvelous ideas and inventions must be omitted. However, having done my first LC experiment nearly 50 years ago [1961[173]], I benefit from practicing LC through nearly all of the second half of its first century. I hope my long perspective has allowed me to make apparent to you the recurring cycle of revolution and evolution in LC, highlighting how insightful prophecy and technical achievement have marked revolutionary milestones on the evolutionary path LC continues to travel from primitive toward ultimate performance.

I still keep handy on my bookshelf, Professors Louis and Mary Fieser's classic text, *Organic Chemistry*.[174] It was remarkable, not just for its clarity and scope, but also for its pioneering demonstration of the Fiesers' belief that young students should know and be inspired by *connections*—by the real people and history behind the science on its pages. Learning all those 'name' reactions is a legendary hurdle that severely challenges all first-year organic chemistry and pre-med students. The Fiesers provided a biographical footnote for each name, thereby establishing their professional pedigree and place in history. In my text, I also make liberal use of annotations that include the titles of each citation and/or relate interesting information about people, places, and things. My hope is that this will not only give you a better sense of the origin of ideas but also stimulate your curiosity, encouraging you to interweave many multidisciplinary paths of inquiry that ultimately expand your LC knowledge.

In the same spirit, you will also note that I have quoted the words of late scientists and living colleagues. It is important to record history while we are still able to remember it accurately! I have been indeed fortunate to have known personally and/or worked side-by-side with so many pioneers in the modern LC community. I have continued to marvel at the manifold *connections* in my life.

Silica is part of my family heritage. About 1850, my late mother's maternal uncle Elliot Robley bought land near Mapleton, Pennsylvania. There he found clay on his island in the river and sand in his nearby ridge.[175] He used the clay to make the bricks for all the local buildings. He opened his first quarry in 1858 [see Figure E–1] and eventually electrified his sandworks not long after Edison invented the electric power industry. His sand—pure silicon dioxide, white as snow—was fused and cast in nearby Corning, New York, to form the mirror blank for the 200-inch Hale telescope on Mt. Palomar. As children we treasured small bags of Juniata River valley sand that provided a stark contrast to the gray Long Island beach sand in our sandbox. Today, the U.S. Silica Company owns and operates the Mapleton Depot sandworks, and its rare output is highly prized by the semiconductor industry as a source of pure silicon.[176]

Figure E–1: My great uncle Elliot Robley's first sand quarry. In 1870 he installed equipment to crush the rock and transported the sand in a stream of water down a quarter-mile-long wooden trough [cleaning it at the same time] to the Pennsylvania Railroad stop where it was dried in equipment he designed and loaded onto hopper cars for transport to the glass making factories in Pittsburgh. [Public domain image from Reference 175, digitized by Google.]

In graduate school, one of my favorite professors, Roy A. Olofson, also a history-of-chemistry buff whose Ph.D. mentor was Prof. Robert Woodward, used to come to class wearing Harvard chemistry department commemorative sweatshirts, including one bearing the image of Louis Fieser, and another celebrating the then recently delineated Woodward-Hoffman orbital symmetry rules. Prof. Olofson spiced his advanced organic synthesis lectures with stories of his Harvard mentors and of the hierarchy of the world's greatest synthetic chemists. In August 2007, I had the opportunity to meet Cornell U. Professor Emeritus, poet, and Nobel laureate Roald Hoffman once again. When I reminded him that I first met him at his series of lectures at Penn State in the late 1960s, he remarked, "That was very early in my career." Trying to recall who he remembered from that visit, he first surfaced the name of Prof. Olofson [who had been his host] and then said something very profound to me. Speaking slowly and thoughtfully, Prof. Hoffman stated, "Roy Olofson—he had very high standards!" My response was immediate: "I was very fortunate. All my teachers had very high standards. I learned from the best!"

Roy Olofson was the one who had recommended that I join Prof. Gordon Hamilton's nascent research group at Penn State. "Why?," I asked. "Because," he stated simply, "Gordon is the most brilliant chemist I know!" [From a small town in Canada, Gordon also did his Ph.D. at Harvard, with the late Prof. Frank Westheimer.] My research in oxidation reaction mechanisms led me to a postdoctoral position with a pioneer in steroid biochemistry, Prof. Seymour Lieberman at Columbia. Seymour had done his doctoral thesis with none other than Louis Fieser. [Together they used to examine urine sample containers in hospital labs all over New York City looking for fortuitously precipitated crystals that might allow them to isolate and identify new steroid metabolites.] Throughout all these years, I continued to develop innovative LC and TLC methods and learn new techniques such as partition LC. During my first year at Columbia, I received a memorable phone call from James Logan Waters and became his customer—and ultimately his colleague and friend. It was my pioneering steroid HPLC separations and ALC-100 instrument modifications that drew the notice of Waters chemists and engineers and led me to join their young company in July 1974.

My greatest joy has been working with, mentoring, and learning from young people. This project is a case in point. In September 2008, Dr. Uwe Neue and I were honored by being selected as members of the inaugural group of eight Waters Corporate Fellows. I have known Uwe since he was a graduate student in Saarbrücken, having been introduced to him by his professor and Waters collaborator István Halász when they both attended an internal training course on Prep LC I presented in Königstein, Germany, in early 1976. I am honored to be Uwe's colleague and to have the benefit of his wise counsel and valuable advice on this volume, as well as all matters chromatographic.[177] Our minds are complementary: Uwe, being a physical chemist, dreams of mathematical equations; as a bioorganic chemist, I visualize entities in spatial relationships and mechanisms.

Equally wonderful and inspirational is my collaboration on this project with three young people trained in disciplines other than chemistry: Ian Hanslope, an exceptional graphic designer and Scouser, formerly with Phase Separations Ltd., who adds elegance with ease to complicated catalogs and detailed documents; Ekaterini Kakouros, a talented designer, relatively new to our team, who tackled this daunting project with enthusiasm and alacrity; and Dawn Maheu, a skilled teacher, now a technical marketing communications specialist whose extraordinary work ethic, powers of observation, and attention to detail—traits that would have served her well in a research career—assure the excellence of our volume. *Connections* are so pivotal—I pray that you are as blessed as I have been to work with, learn from, and be inspired by the best from all generations!

Pat McDonald
Milford, Massachusetts, June 2009

Annotations

Page 5

1   J.C. Arsenault and P.D. McDonald, *A Beginner's Guide to Liquid Chromatgraphy*, Waters, Milford [2009]; order Part No. **715001531** on waters.com .

Page 7

2   Opening lines in the book published from his Warsaw thesis: M. S. Tswett, *Khromofillii v Rastitel'nom i Zhivotnom Mire [Chlorophylls in the Plant and Animal World]*, Izd. Karbasnikov, Warsaw, 1910; for a partial English translation, see pp. 35–79 in: V.G. Berezkin, Ed., *Chromatographic Adsorption Analysis – Selected Works of M.S. Tswett*, Ellis Horwood, Chichester [1990].

3   (a) M. S. Tswett, "*[Physical chemical studies on chlorophyll adsorptions]*," *Ber. Deutsch. bot. Ges.* **24**: 316–323 [1906]; (b) M.S. Tswett, „*[Adsorption analysis and chromatographic method. Application to the chemistry of chlorophyll]*," *Ber. Deutsch. bot. Ges.* **24**: 384–393 [1906]; for a translation of excerpts, see: http://web.lemoyne.edu/~GIUNTA/tswett.html .

4   V.R. Meyer, "*Michael Tswett's Columns: Facts and Speculations*," *Chromatographia* **34(5–8)**: 342–346 [1992].

5   It is curious that Tswett's work was not cited, by either the examiner or the inventors, as prior art in the application for "*Flash chromatography*," *U.S. Patent* 4,293,422 [1981]. As his contemporary George Santayana wrote about the same time as Tswett's first papers were published, "Those who cannot remember the past are condemned to repeat it." [*The Life of Reason* **Vol. 1**, *Reason in Common Sense*, 1905].

Page 8

6   (a) K. Sakodynskii, K. Chmutov, "*M.S. Tswett and Chromatography (To the 100th Anniversary of M.S. Tswett's Birthday)*," *Chromatographia* **5(8)**: 471–476 [1972]; (b) E.M. Senchenkova, *Michael Tswett, the Creator of Chromatography*, V.A. Davankov and L.S. Ettre, Eds. of English Translation, Russian Academy of Sciences, Scientific Council on Adsorption and Chromatography, Moscow [2003].

7   Tswett earned his first Ph.D. in Geneva in 1896. He did a second Ph.D. thesis in Warsaw [then a city in Russia]. See: L.S. Ettre and K.I. Sakodynskii, "*M.S. Tswett and the Discovery of Chromatography II: Completion of the Development of Chromatography (1903–1910)*," *Chromatographia* **35(5/6)**: 329–338 [1993] and references cited therein.

8   R. Willstätter, A. Stoll, *[Studies on Chlorophyll]*, Springer, Berlin [1913]; R. Willstätter, *[Studies on Enzymes]*, Vol. I, Springer, Berlin [1928], p. 295.

9   R. Kuhn, A. Winterstein, and E. Lederer, "*[Contribution to the Knowledge of Xanthophyll]*," *Hoppe-Seyler's Z. Physiol. Chem.* **197**: 158 [1931].

10 E. Lederer, "*La Renaissance de la Méthode Chromatographique de M. Tswett en 1931*," *J. Chromatogr.* **73**: 361–366 [1973] and references cited therein.

11 Brockmann later created the standard activity grade scale, based upon dye retention *vs.* water content, for chromatographic alumina packings. See: H. Brockmann and H. Schodder, "*[Aluminum Oxide with Differential Capacity for Chromatographic Adsorption]*," *Chem. Ber.* **74(1)**: 73–78 [1941].

12 In a plenary lecture at the 1947 IUPAC Congress, Karrer stated, "*No other discovery has exerted as great an influence and widened the field of investigation of the organic chemist as much as Tswett's chromatographic adsorption analysis.*"

13 Speaking in southern California at a local section meeting of the American Chemical Society on November 3, 1950, Zechmeister argued: "*Recently chromatography has become so popular that the English language has been enriched by a new noun – 'the chromatographer'. I would protest against such a label. In research, chromatography should be considered first of all a tool like, e.g., fractional distillation; and those of our colleagues who have achieved success by using distillation methods should certainly not be named 'distillers'.*"

14 This work, supported by Nestlé, was the foundation for the development of the first instant coffee! Personal communication to P. D. McDonald from his postdoctoral mentor, Prof. Seymour Lieberman [see Vitae], 1970. Seymour's Ph.D. and postdoctoral mentors, in turn, were Prof. Louis Fieser, Harvard U., Cambridge, and Prof. Tadeus Reichstein, ETH, Zürich.

15 T. Reichstein and C.W. Shoppee, "*Chromatography of Steroids and Other Colourless Substances by the Method of Fractional Elution*," *Discuss. Faraday Soc.* **7**: 305–311 [1949].

16 Biographies of all these scientists at http://nobelprize.org are fascinating to read.

17 "*In the fields of observation, chance favors only the mind that is prepared.*" Louis Pasteur (1822–1895) [quoted by R. Vallery-Radot, 1927]. In a similar vein, Paul Flory echoed in his 1981 Perkin Medal Award address, "*Happenstance usually plays a part, to be sure, but there is much more to invention than the popular notion of a bolt out of the blue. Knowledge in depth and in breadth are virtual prerequisites. Unless the mind is thoroughly charged beforehand, the proverbial spark of genius, if it should manifest itself, probably will find nothing to ignite.*"

18 Archer J.P. Martin, "*The development of partition chromatography*," Nobel Lecture, Dec. 12, 1952; full text available online at: http://nobelprize.org/ nobel_prizes/chemistry/laureates/1952/martin-lecture.pdf .

19 A.J.P.Martin, Chap. 31 in: L.S. Ettre and A. Zlatkis, Eds. "*75 Years of Chromatography—a Historical Dialogue*," *J. Chromatogr. Lib.* **17**: p. 286 [1979].

20 At about the same time, but an ocean away, American inventor Lyman Craig at the Rockefeller U. was working on amino acid separations. He built a train of several hundred tubes that could perform an unattended series of sequential liquid-liquid extractions—the Craig Countercurrent Distribution Apparatus. Unlike Martin, he never envisioned partition as a chromatographic technique.

Page 9

21 *J. Chromatogr. Library* **17**: 287 [1979].

22 *Biochem. J.* **35**:1358–1368 [1941].

23 Reference 22, p. 1359.

24 A.T. James and A.J.P. Martin, "*Gas-liquid partition chromatography: the separation and microestimation of volatile fatty acids from formic acid to dodecanoic acid*," *Biochem. J.* **50**: 679–690 [1952].

25 Reference 22, p. 1363.

26 J.C. Giddings, *Unified Separation Science*, Wiley, New York [1991], p. 289; NOTE: This book was created over a 21-year period to fulfill Giddings' goal of showing the commonality of mass transport phenomena that underlie all separation processes. He died of cancer five years after its publication.

27 (a) J.C. Giddings, *Dynamics of Chromatography, Part 1, Principles and Theory* Marcel Dekker, New York [1965]; NOTE: Part 2 was never written; (b) B.L. Karger, L.R. Snyder, and C. Horváth, "*An Introduction to Separation Science*," Wiley, New York [1973]; (c) J.H. Knox, Ed., "*High Performance Liquid Chromatography*," Edinburgh University Press [1978]; (d) U.D. Neue, "*HPLC Columns: Theory, Technology, and Practice*," Wiley-VCH, New York [1997].

28 P.D. McDonald and B.A. Bidlingmeyer, "*Strategies for Successful Preparative Liquid Chromatography*," *J. Chromatogr. Lib.* **38**, 1–103 [1987]; see esp. pp. 11–28.

Page 10

29 By convention, in this ratio, the retention factor for the later eluting compound is placed in the numerator so that α is always greater than one.

30 This simulated chromatogram was drawn using a spreadsheet created by Uwe Neue. Parameters: plate count = 5000; respective peak areas: 300, 3000, 6000; respective k values: 0, 4, and 5. For instructions on how to do this yourself, see: U. Neue, "*Simulation of Chromatograms*," Chap. 4, pp. 83–103 in: H-J. Kuss and S. Kromidas, Eds., *Quantification in LC and GC: A Practical Guide to Good Chromatographic Data*, Wiley-VCH, Wenheim [2009].

31 See p. 20 and p. 24 in reference 28.

32 K. Miyamoto, T. Hara, H. Kobayashi, H. Morisaka, D. Tokuda, K. Horie, K. Koduki, S. Makino, O. Nunez, C. Yang, T. Kawabe, T. Ikegami, H. Takubo, Y. Ishihama, and N. Tanaka, "*High-Efficiency Liquid Chromatographic Separation Utilizing Long Monolithic Silica Capillary Columns*," *Anal. Chem.* **80(22)**: 8741–8750 [2009]; see also reference 107.

Page 11

33 An injection volume of 5 µL of a sample solution with an analyte [MW 300] concentration of 1 ng/µL contains about 10^{13} molecules of that compound! [Remember: one mole [in this case, 300 g] contains 6.023 x 10^{23} molecules (Avogadro's number, ironically symbolized by N in chemical calculations—not to be confused with N = plate count).]

34 This assumption is reasonable for an isocratic LC system. However, in gradient LC, band widths are artificially compressed as the elution strength of the mobile phase increases with time and distance traveled. Therefore, any calculation of N using parameters gleaned from a gradient chromatogram is meaningless. Likewise, if a peak shape on a chromatogram is not Gaussian, then methods used to calculate N shown in Figure 5–1 are not valid.

35 NOTE: Modern LC data systems use an internal clock controlled by a microprocessor to measure retention and peak width in units of time. As stated earlier in the discussion related to Figure 4–1, elution volume may be converted to units of time if the volume flow rate is constant. Sophisticated design is required to build mobile phase delivery systems that can maintain a constant volume flow rate so that calculations of LC system and separation parameters may be accurately made using variables expressed in time units. This task becomes especially difficult when operating at high pressure with mixtures of compressible fluids whose viscosities change with temperature and gradient composition.

36 J.J. van Deemter, F.J. Zuiderweg, and A. Klinkenberg, "*Longitudinal diffusion and resistance to mass transfer as causes of nonideality in chromatography*," *Chem. Eng. Sci.* **5**: 271–289 [1956]. This work was presented at a 1956 Gordon Conference while Van Deemter was working at Shell Houston [1955–1957].

Page 12

37 See reference 27c, p. 11.

38 TIDE®, introduced by the Proctor and Gamble Company in 1945. Its utility in GC separations was discovered in 1960 [A.W. Decora and G.U. Dinneen, "*Gas-Liquid Chromatography of Pyridines Using a New Solid Support*," *Anal. Chem.* **32(2)**: 164–169 [1960]].

39 Charles Pidacks, resident LC expert at Lederle Laboratories, Pearl River, New York, in the 1940s–1960s, standardized a technique for LLPC. (See: M.J. Weiss, R.E. Schaub, G.R. Allen, Jr., J. F. Poletto, C. Pidacks, R.B. Conrow, and C.J. Coscia, "*The Formation of Steroid Enolate Anions by Reductive Procedures*," *Tetrahedron* **20**: 357–372 [1964]) Charlie participated on the teams that developed the seminal antibiotics aureomycin®, mitomycin and puromycin and holds patents on the LC procedures used in their production (See, *e.g.*: C. Pidacks and E.E. Starbird, "*Purifying Aureomycin by Chromatographic Adsorption*," *U.S. Patent 2,586,766* [1952]). His technician at Lederle, Pentti 'Finn' Siiteri, introduced the Pidacks partition technique to the laboratory of Prof. Seymour Lieberman, Columbia U. [see reference 14], and described it in his 1963 Ph.D. thesis in steroid biochemistry. There it was refined to a high art, including reversed-phase and ion-pairing modes of separation. See: P.D. McDonald, *Adv. Chromatogr.* **42**: pp. 328–329 [2003] and references 8–13 cited therein. From there, the Pidacks procedure spread to Europe: see P. Robel, R. Emiliozzi, and É.-É. Baulieu, "*Studies on Testosterone Metabolism*," *J. Biol. Chem.* **241(1)**: 20–29 [1966]. Dr. Émile-Étienne Baulieu was a visiting scientist in Lieberman's laboratory from 1961–1962. His work on the progesterone receptor led him to suggest to Roussel-Uclaf the development of an antiprogesterone drug [RU486 or Mifepristone].

40 Alumina was widely available as a product of the process invented by Karl Bayer in 1887 while working in St. Petersburg to produce dye mordants for the textile industry. Coupled with the Hall-Héroult electrolytic process developed in 1886 that could convert alumina to aluminum metal, the Bayer process marked the beginning of modern hydrometallurgy.

Page 13

41 By convention, both mesh and particle diameter ranges are written with the smaller number first. To avoid confusion, however, keep in mind that the lower mesh size number indicates a larger particle diameter, and *vice versa*.

42 R. Consden, A.H. Gordon, and A.J.P. Martin, "*Qualitative analysis of proteins: a partition chromatographic method using paper*," *Biochem. J.* **35**: 224–232 [1944].

43 E. Stahl, "*Thin-layer chromatography*," *Pharmazie* **11**: 633–7 [1956]; "*Thin-layer chromatography. II: standardization, visualization, documentation, and application*," *Chem. Ztg.* **82**: 323–9 [1958].

44 (a) E. Stahl, "*This Week's Citation Classic*," *Current Contents* **52**(Dec. 24–31): 226 [1979]; (b) E. Stahl, "*Twenty Years of Thin-Layer Chromatography. A Report on Work with Observations and Future Prospects*," *J. Chromatogr.* **165**: 59–73 [1979].

45 E. Stahl, "*Dünnschicht-Chromatographie: Ein Laboratoriumshandbuch [Thin-layer Chromatography: A Laboratory Handbook]*," Springer, Berlin [First Ed. 1962; English Translation 1965; Second Edition 1967].

46 TLC is really a double gradient separation as both the flow rate and the mobile phase composition change during the development of the separation. The solvent front moves more slowly the further it proceeds up the plate. The dry, active stationary phase layer selectively adsorbs the more polar constituent(s) in the mobile phase, thereby locally changing its composition and weakening its strength. This makes it tricky to reproduce and/or to scale up a TLC separation to LC or HPLC; *e.g.*, see: L.R. Snyder, "*R_F Values in Thin-Layer Chromatography on Alumina and Silica*," *Adv. Chromatogr.* **4**: 3–46 [1967]; S. Hara, "*Use of Thin-Layer Chromatographic Systems in High-Performance Liquid Chromatographic Separations. Procedure for Systematization and Design of the Separation Process in Synthetic Chemistry*," *J. Chromatogr.* **137**(1): 41-52 [1977]. NOTE: Prof. Shoji Hara, now long retired from the Tokyo College of Pharmacy, was not only a good chromatographer, but also an entertaining magician, often the center of attention at chromatography symposia dinner tables!

47 (a) S. Moore, D.H. Spackman, and W.H. Stein, "*Chromatography of Amino Acids on Sulfonated Polystyrene Resins*," *Anal. Chem.* **30**(7): 1185–1190 [1958]; (b) D.H. Spackman, W.H. Stein, and S. Moore, "*Automatic Recording Apparatus for Use in the Chromatography of Amino Acids*," *Anal. Chem.* **30**(7): 1190–1206 [1958]; (c) L.S. Ettre and C.W. Gehrke, "*The Development of the Amino Acid Analyzer*," *LC-GC* **24**(4): 390–400 [2006].

48 S. Moore and W.H. Stein, "*The Chemical Structures of Pancreatic Ribonuclease and Deoxyribonuclease*," Nobel Lecture, December 11, 1972; pdf available at: http://nobelprize.org/nobel_prizes/chemistry/laureates/1972/moore-lecture.html .

Page 14

49 (a) P.D. McDonald, "*James Waters and His Liquid Chromatography People: a Personal Perspective*," *Waters Whitepaper* **WA62008***: 20 pp [2006]; (b) P.D. McDonald, "*Waters Corporation: Fifty Years of Innovation in Analysis and Purification*," *Chemical Heritage* **26**(2): 32–37 [2008]; **WA60206***; (c) Y. Hager, "*Going with the flow*," *Chemistry World* **5**(9): 70–74 [2008]; **WA63958***; (d) James L. Waters, interview by Arnold Thackray and Arthur Daemmrich in Framingham, Massachusetts, 21 August 2002 (Philadelphia: Chemical Heritage Foundation, *Oral History Transcript #0262*); (e) L.S. Ettre, "*Jim Waters: the Development of GPC and the First HPLC Instruments*," *LC-GC* **23**(8): 752–761 [2005].
*NOTE: Search for these codes at waters.com to view corresponding PDFs.

Page 15

50 J.C. Moore, "*Separation of Large Polymer Molecules in Solution*," *U.S. Patent* 3,326,875 [filed Jan. 1963; issued June 1967].

51 J. Porath and P. Flodin, "*Gel Filtration: A Method for Desalting and Group Separation*," *Nature* **183**: 1657-1659 [1959].

52 J.F.K. Huber, Chap. 17 in: *J. Chromatogr. Library* **17**: 159–166 [1979].

53 (a) C. Horváth, Chap. 16 in: *J. Chromatogr. Library* **17**: 151–158 [1979]; (b) C. Horváth, "*My focus on chromatography over 40 years*," *J. Chromatogr. Library* **64**: 238–247 [2001].

54 Prof. Lipsky invited Csaba to join his new GC/MS laboratory, funded by NIH and NASA, in preparation for the analysis of lunar rock samples for biological molecules that might indicate the possibility of past, present, or future life on Earth's moon. Fortuitously, the space missions were several years away, so, while waiting, Csaba sought and received Sandy's blessing to build a liquid chromatograph to process biological samples.[53b]

55 I. Halász and C. Horváth, "*Micro Beads Coated with a Porous Thin Layer as Column Packing in Gas Chromatography*," *Anal. Chem.* **36**: 1178–1186 [1964].

56 C.G. Horváth, B.A. Preiss, and S.R. Lipsky, "*Fast Liquid Chromatography: An Investigation of Operating Parameters and the Separation of Nucleotides on Pellicular Ion Exchangers*," *Anal. Chem.* **39**(121): 1422–1428 [1967].

57 (a) J.J. Kirkland, Ed., *Modern Practice of Liquid Chromatography*, Wiley-Interscience, New York [1971]; (b) L.R. Snyder and J.J. Kirkland, *Introduction to Modern Liquid Chromatography*, Wiley-Interscience, New York [1974]. For many years, Jack and Lloyd taught the highly successful American Chemical Society short course on HPLC.

58 J.J. Kirkland, Chap. 23 in: *J. Chromatogr. Library* **17**: 209–217 [1979].

59 (a) J.J. Kirkland, "*Superficially Porous Supports for Chromatography*," *U.S. Patent* 3,505,785 [1970] and references therein, *e.g.*, R.K. Iler, "*Articles Coated with Colloidal Particle Multilayers*," *Canadian Patent* 729,581 [1966]; (b) J.J. Kirkland, "*Controlled Surface Porosity Supports for High Speed Gas and Liquid Chromatography*," *Anal. Chem.* **41**(1): 218–220 [1969]; (c) NOTE: A modern version of this technology was introduced in 2007 [see: J.J. Kirkland, T.J. Langlois, and J.J. DeStefano, "*New Fused-Core® Particles for Very Fast HPLC Separations*," *Lecture*, Pittcon 2007; PDF available: http://www.advanced-materials-tech.com/pittcon.pdf]. A porous outer layer was formed on a solid, non-porous, 1.7-μm-diameter silica core, creating a 2.7-μm-diameter particle. This material has commonly been described in the literature as *superficially porous*. However, this is a misnomer as, in fact, 75% of the total geometric volume of such a particle is porous; the solid core occupies only 25% of the particle by volume. With a surface area of 150 m²/g and 90-Å average pore diameter, such particles behave, as expected, like totally porous particles of about the same size. As with the

traditional pellicular packings, the solid core increases the particle's density and enables packing a column bed with good efficiency. When separations are run on a system optimized for minimal extra-column band spreading, well-packed columns containing smaller particles exhibit higher efficiency than do corresponding columns filled with larger particles, no matter whether those particles are totally or mostly porous. An important consideration with any pellicular or porous particle is the chemical nature of the silica substrate; see, *e.g.*, K.J. Fountain, J. Xu, Z. Yin, P.C. Iraneta, and D.M. Diehl, *"Comparison of Fully and Superficially Porous Particle Columns for the Analysis of Basic Compounds,"* Waters Application Note 5 pp. [2008]; PDF available, search for **72002825en** on waters.com. A final caution: do not equate the full linear thickness of the porous pellicule or the radius of a totally porous particle with the diffusion distance. Only a small fraction of the total porosity of a particle is accessible to analyte molecules within the context of an LC separation. For a discussion of the often misconstrued role of particle morphology in LC separations, see reference 171.

Page 16

60 J.J. Kirkland, *"A High-Performance Ultraviolet Photometric Detector for Use with Efficient Liquid Chromatographic Columns,"* Anal. Chem. **40(2)**: 391–396 [1968].

61 I. Halász and C. Horváth, *"Gas Chromatographic Separating Material,"* U.S. Patent 3,340,085 [1967]; this patent was assigned to the Perkin-Elmer Corporation.

62 In the abstract to his seminal presentation at the Fifth International Symposium on Advances in Chromatography, Las Vegas, 20–23 January 1969, Prof. Huber used the phrase, *"high performance liquid chromatograph"*—but not its corresponding acronym. His lecture was published a month later: J.F.K. Huber, *"High Efficiency, High Speed Liquid Chromatography in Columns,"* J. Chromatogr. Sci. **7**: 85–90 [1969].

63 Personal communication from C. Horváth to P.D. McDonald. In reference 53a, Csaba tells the story a bit differently. In his lecture, he introduced a new adjective he had coined [after consulting both Greek and Latin dictionaries] to describe elution at constant eluent strength: *isocratic*. Within the hour, an enterprising vendor in his exhibition booth, to disguise the lack of gradient capability in his newest LC instrument, proclaimed: *"Our system is specially designed for isocratic elution!"*

64 L.S. Ettre, *"Csaba Horváth and the Development of the First Modern High Performance Liquid Chromatograph,"* LC-GC **23(8)**: 752–761 [2005].

65 R.B. Woodward, *"The Total Synthesis of Vitamin B₁₂,"* Pure Appl. Chem. **333**: 145-177 [1973]; see esp. pp. 165–166.

66 Polish-born Prof. Albert Zlatkis, U. Houston, a GC pioneer, in 1963 organized the first in an annual series of International Symposia on Advances in Chromatography, bringing together the best and brightest chromatography practitioners. Without the help of a committee, he continued these conclaves for more than two decades; hence, they became known familiarly as the *Zlatkis meetings*. See: L.S. Ettre, *"Sixtieth Birthday of Albert Zlatkis,"* Chromatographia **18(5)**: 233 [1984].

Page 17

67 Dick joined Waters Associates in late 1972 after finishing his Ph.D. thesis [*The Measurement of Formation Constants of Hydrogen-Bonded Complexes by Gas-Liquid Chromatography*] with then Associate Professor Barry Karger at Northeastern University in 1971, a postdoctoral year with the late Roland Frei at Dalhousie University, Halifax, Nova Scotia, and a short stint at Sandoz-Wander, Hanover, New Jersey [A.F. Michaelis, D.W. Cornish, and R. Vivilecchia, *"High Pressure Liquid Chromatography,"* J. Pharm. Sci. **62(9)**: 1399–1416]. With help from Norma Thimot and the late Richard Cotter, Dick was the first to synthesize octadecyldimethylchlorosilane from 1-octadecene and dimethylchlorosilane using hexachloroplatinic acid as a catalyst.* Five years later, when Waters Associates contracted an outside company to manufacture this silane, they permitted its sale to other column manufacturers. In early 1973, Dick V. and Dick Cotter visited the Saarbrücken laboratories of Prof. István Halász to gain a better understanding of slurry-packing methods for small particles [U. Neue, personal recollection]. Dick left Waters in 1979 to rejoin Sandoz; he retired from Novartis in April 2008 as Executive Director of Analytical Research and Development.

*This reaction had been invented by Dr. John Speier at Dow Corning Corporation (see: J.L. Speier, J.A. Webster, and G.H. Barnes, *"The Addition of Silicon Hydrides to Olefinic Double Bonds. Part II. The Use of Group VIII Metal Catalysts,"* J. Am. Chem. Soc. **79(4)**: 974–979 [1957]). Speier received the *Frederic Stanley Kipping Award in Silicon Chemistry* at the National ACS Meeting in Boston, 1990; in his biographical, celebratory remarks on that occasion, the research director of Dow Corning noted that 75% of their revenue was based upon the products of reactions invented by Speier! [P.D.McD., personal recollection] In 1978, Industrial Research/Development magazine voted Speier *Scientist of the Year*. (B. Sharp, *Industrial Research* **20**: 65-66 [1978]; R.R. Jones, *Ibid.* **20**: 9 [1978]) In his accompanying article in the same issue, Speier reviewed some of the significant chemistry he had developed in his career. He noted by way of a diagram that the reaction of a chloro- or alkoxysilane with a silanol on a silica surface to form a siloxane bond is *reversible*. This fact—that a bonded phase on silica, while *thermally* stable, was *hydrolytically* less stable—was overlooked by many early HPLC practitioners. (J.L. Speier, *Ibid.* **20**: 131-134 [1978]).

Page 18

68 Details of the pump may be found in: L. Abrahams, B.M. Hutchins, and J.L. Waters, *"Novel Pumping Apparatus,"* U.S. Patents 3,855,129 [filed March 1972; issued December 1974] and 3,981,620 [1976]. The innovative check valves were disclosed in: L. Abrahams and B.M. Hutchins, *"Check Valve and System Containing Same,"* U.S. Patent 3,810,716 [1974].

69 Jorgenson and his students modified an M6000 pump to do UHPLC gradient separations at 15,000 psi [1000 bar]: L. Tolley, J.W. Jorgenson, and M. A. Moseley, *"Very High Pressure Gradient LC/MS/MS,"* Anal. Chem. **73(13)**: 2985–2991 [2001].

70   Details may be found in: L. Abrahams and B.M. Hutchins, *"Novel Injector Mechanism,"* U.S. Patent 3,916,692 [1975]. The novel valve is the subject of two additional patents: L. Abrahams, *"Novel Gasket and Valve Comprising Same,"* U.S. Patent 3,918,495 [1975]; and L. Abrahams, *"Novel Gasket and Flow Cell Comprising Same,"* U.S. Patent 4,027,983 [1977]. The latter discloses how the same technology has been applied to optical detector flow cells. A further extension is described in the caption to Figure 7–5.

71   D.R. Friswell and L.J. Finn, *"Novel Seal and Apparatus Including Same,"* U.S. Patent 4,094,195 [1978].

72   For example, a new material was invented to fabricate superior, more chemically inert, and longer-lasting piston seals: L. Abrahams and T.P.J. Izod, *"Wear-Resistant Article,"* U.S. Patent 4,333,977 [1982].

73   P.D. McDonald, personal communications with colleagues, both at Waters and at many international LC symposia, over a period of 35 years.

74   The first µBondapak C_{18} columns were made using a monomeric layer of octadecyltrichlorosilane. (See: R.V. Vivilecchia, R.L. Cotter, R.J. Limpert, N.Z. Thimot , and J. N. Little, *"Considerations of Small Particles in Different Modes of Liquid* Chromatography," *J. Chromatogr.* 99: 407–424 [1974]) A year later, in the summer of 1974, faced with a rising demand for columns and a short supply of silica, I suggested to Dick that a type of silica I had been investigating for its desirable properties for preparative LC applications might also be used to advantage in µPorasil and µBondapak columns. This change was made. After subjecting the new silica to proprietary grinding, classification, chemical treatment, and bonding processes [using Dick's new monofunctional octadecyldimethylchlorosilane, see reference 67], unique to this day, the new, improved µBondapak C_{18} column, re-introduced early in the fall of 1974, became the best-selling and most widely referenced HPLC column in history and is still sold today.

75   M. Holdoway, personal letter to P.D. McDonald, January 2009. Mike had described his new silica on a 1974 visit to Waters Associates in Milford. Mike retired from Phase Separations in 1984 and founded a new company, Exmere, where he developed an elegant continuous process that could control any two of three primary variables—pore size, pore size distribution, and pore volume—in small-particle spherical silica. This process was licensed to several companies; one of these, Alltech, purchased Exmere in 1994. Mike then retired with his wife Maureen to the English seacoast. In 1996, Phase Separations became a wholly owned subsidiary of Waters Corporation. Mike takes pride in knowing that much of the world's spherical silica, under many brand names, can be traced back to technology that he had developed!

76   The Saint-Gobain company was founded in the Faubourg Saint-Antoine section of Paris in 1665 to ensure the glory of King Louis XIV by achieving independence from the Venetian technical and commercial monopoly in making mirrors and glass. Artisans, who had emigrated from Venice to Saint-Gobain [followed later by assassins sent to dispatch them], invented a novel high-temperature glass-casting process to fabricate the mirrors and windows for one of the most famous rooms in the world, the Hall of Mirrors in the Palace of Versailles. Two modern-day Saint-Gobain projects of note were I.M.

Pei's glass pyramid entrance to the Louvre Museum [location of the ultimate scene in the best-selling Dan Brown novel, *The Da Vinci Code*] and the Glass Bridge Skywalk overhanging the western rim of the Grand Canyon. The Péchiney-Saint-Gobain chemical manufacturing subsidiary was acquired by Rhône-Poulenc in 1969.

77   M. Le Page, R. Beau, and J. Duchêne, *"Grains de silice poreux à texture définie,"* French Patent 1,473,240 [1967]; M. Le Page and A. de Vries, *"Remplissages déstines aux colonnes de séparation par chromatographie,"* French Patent 1,475,929 [1967]; M. Le Page and A. de Vries, *"Chromatographic Separation with Porous Silica,"* U.S. Patent 3,677,938 [1972].

78   A.J. de Vries, M. LePage, R. Beau, and C.L. Guillemin, *"Evaluation of Porous Silica Beads as a New Packing Material for Chromatographic Columns. Application in Gel Permeation Chromatography,"* Anal. Chem. **39(8)**: 935–939 [1967].

79   J. Vermont, M. Deleuil, A.J. de Vries, and C.L. Guillemin, *"Modern Liquid Chromatography on Spherosil,"* Anal. Chem. **47(8)**: 1329–1337 [1975].

80   In his classic treatise, Lloyd Snyder states that practical LC adsorbents have pore diameters > 20 Å. See pp. 61–62 and 105–107 in: L.R. Snyder, *Principles of Adsorption Chromatography. The Separation of Nonionic Organic Compounds*, Marcel Dekker, New York [1968].

81   I. Sebastian and I. Halász, *"Chemically Bonded Phases in Chromatography,"* Adv. Chrom. **14**: 75–86 [1976; written in 1974].

82   See for example: p. 78 in reference 81 or p. 84 in reference 87a. An ester is the product of the reaction of an acid with an alcohol. An ether is most commonly the product of the intermolecular dehydration of two alcohols. Earlier workers presumably termed a silyl ether as a silyl ester because a silyl alcohol [silanol] is more acidic than an alkyl alcohol. Here, we prefer to abide by rules of organic nomenclature, where names of organosilicon compounds are analogous to those of corresponding carbon compounds.

83   O.-E. Brust, I. Sebastian, and I. Halász, *"Stationäre phasen mit $\equiv Si-N\equiv$ bindung für die flügkeitschromatographie,"* J. Chromatogr. **83**: 15–24 [1973].

84   D.C. Locke, J.T. Schmermund, and B. Banner, *"Bonded stationary phases for chromatography,"* Anal. Chem. **44(1)**: 90–92 [1972].

85   See 27d and subsequent references 87a, 88, and 106.

86   K. Unger, J. Schick-Kalb, *"Poroeses Siliciumdioxid [Spherical porous silicon dioxide – produced from polyalkoxysiloxanes],"* German patent 2,155,281A [filed: November 6, 1971; issued: May 24, 1973].

87   (a) K. K. Unger, *"Porous Silica: Its Properties and Use as a Support in Column Liquid Chromatography,"* J. Chromatogr. Lib **16**: 336 pp [1979]; Four companies supported this book with full-page advertisements for their range of LC packings inside the rear cover: Woelm Pharma, Whatman, Shandon Southern, and E. Merck. (b) In his review of this book (*J. Liq. Chromatogr.* **3(4)**: 611–612 [1980]), P.D. McDonald wrote, *"It takes a good deal of experience and effort to pull together from diverse disciplines the data and*

dogma, culling fact from fiction, fundamental to an interdisciplinary field such as chromatography. ... Professor Unger's valuable contribution to the chromatographic literature can be highly recommended to anyone interested in a better understanding of the science of chromatography."

88 K.K. Unger, R. Skudas, and M.M. Schulte, *"Particle packed columns and monolithic columns in high-performance liquid chromatography—comparison and critical appraisal,"* J. Chromatogr. A **1184**: 393–415 [2008].

89 I. Halász and I. Sebestian, *"Process for Preparing Silica Particles,"* U.S. Patent 3,857,924 [filed: November 2, 1972; issued: December 31, 1974]; an earlier *German Patent* 2,155,045 was filed on November 5, 1971.

90 Personal communication to P. McDonald from U. Neue, April 2009.

91 Capping a brilliant career in silica chemistry, Iler wrote a comprehensive masterwork: *The Chemistry of Silica: Solubility, Polymerization, Colloid and Surface Properties and Biochemistry of Silica*, Wiley-Interscience, New York [1979]. He begins, with keen insight, Chapter 1 on page 3: *"Silica is by far the major component of the earth's crust, yet much remains to be learned of its chemistry and, in particular, its solubility behavior in water. ... As water is a unique liquid, so is amorphous silica a unique solid. They are much alike, both consisting mainly of oxygen atoms with the smaller hydrogen or silicon atoms in the interstices."*

92 (a) E.C. Broge and R.K. Iler, *"Process of Increasing the Size of Unaggregated Silica Particles in an Aqueous Silica Suspension,"* U.S. Patent 2,680,721 [1954] ; (b) R.K. Iler, *"Aqueous Silica Dispersions,"* U.S. Patent 2,956,958 [1960].

93 (a) R.K. Iler and H.J. McQuestion, *"Uniform Oxide Microspheres and a Process for Their Manufacture,"* U.S. Patent 3,855,172 [Filed: April 7, 1972; Issued: December 17, 1974]; (b) *Ibid*. Divisional *U.S. Patent* 4,010,242 [Filed: Nov. 20, 1973; Issued: March 1, 1977]. Note that these patents claim a process that, in addition to silica, can also be used to make porous particles of other metal oxides, such as alumina, zirconia, titania, or combinations thereof.

94 J.J. Kirkland, *"Completely Porous Microspheres for Chromatographic Uses,"* U.S. Patent 3,782,075 [Filed: April 7, 1972; Issued: January 1, 1974]. It is interesting to note that this 'use' patent, though filed on the same day but reviewed by a different examiner, issued nearly a year earlier than Iler's 'process' patent cited in reference 93a.

95 J.J. Kirkland, *"High-Performance Liquid Chromatography with Porous Silica Microspheres,"* J. Chromatogr. Sci. **10**: 593–599 [1972].

96 In early 1976, Paul Raven left Edinburgh to use his LC skills in the Analytical Control Division of Allen & Hanburys, Ltd., a British pharmaceutical maker in Ware that had been absorbed by Glaxo in 1958. At the Ware facility, Allen & Hanburys specialized in infant formulas and foods. They also pioneered the production of cod liver oil and medicated pastilles in Great Britain.

97 In 1972, the Wolfson Foundation, as part of its program *"to link scientific research with industry,"* awarded Knox a grant to initiate this Unit. In 1974, a fruitful collaboration was established with Shandon Southern Products Ltd, and two years later it resulted in a production prototype of a full HPLC instrument. See: http://homepages.ed.ac.uk/prrls02/jhk.html .

98 J.H. Knox and A. Pryde, *"Performance and Selected Applications of a New Range of Chemically Bonded Packing Materials in High-Performance Liquid Chromatography,"* J. Chromatogr. **112**: 171–188 [1975].

99 S.T. Sie and N. van den Hoed, *"Preparation and Performance of High-Efficiency Columns for Liquid Chromatography,"* J. Chromatogr. Sci. **7**: 257–266 [1969].

100 J.J. Kirkland, *"High Speed Liquid-Partition Chromatography with Chemically Bonded Organic Stationary Phases,"* J. Chromatogr. Sci. **9**: 209–214 [1971].

101 R.E. Majors, *"High Performance Liquid Chromatography on Small Particle Silica Gel,"* Anal. Chem. **44(11)**: 1722–1726 [1972].

102 W. Strubert, *"Herstellung von Hochleistungssäulen für die schnelle Flüssigkeits-Chromatographie [Preparation of High-Efficiency Columns for High Speed Liquid Chromatography],"* Chromatographia **6(1)**: 50–52 [1973].

103 See Chapter 5 in reference 27d.

104 In addition to reference 85, please read: U.D. Neue, *"Silica Gel and Its Derivatization for Liquid Chromatography,"* in: *Encyclopedia of Analytical Chemistry* R.A. Meyers (Ed.), pp. 11450–11472, Wiley, New York [2000]; revised, updated version in press [2009].

105 A pioneer in the use of small-bore columns [1 mm-i.d.], and the necessary instrument modifications for minimal band spreading and maximum detector sensitivity, was Raymond P.W. Scott. He carried out separations on long, slow [5 µm silica, 10 x 1 m., achieving 1,000,000 plates in 36 hours] and short, fast columns [20 µm silica, 10 cm, 30 seconds] at Hoffman-La Roche, Nutley, New Jersey, and later as Director of the Applied Research Laboratory, Perkin-Elmer Corporation, Norwalk, Connecticut. See, for example: (a) R.P.W. Scott and P. Kucera, *"Mode of Operation and Performance Characteristics of Microbore Columns for Use in Liquid Chromatography,"* J. Chromatogr. **169**: 51–72 [1979]; (b) R.P.W. Scott, P. Kucera, and M. Munroe, *"Use of Microbore Columns for Rapid Liquid Chromatographic Separations,"* J. Chromatogr. **186**: 475–487 [1979]; (c) R.P.W. Scott, Ed., *Small Bore Liquid Chromatography Columns: Their Properties and Uses*, John Wiley, New York [1984] and references therein.

Page 23

106 L. Abrahams and M.A. Russo, *"Chromatographic Column with Improved End Fittings,"* U.S. Patent 3,026,803 [filed: December 24, 1975; issued: May 31, 1977].

107 L. Abrahams, *"Chromatography Tube,"* U.S. Patent 4,070,285 [filed: December 24, 1975; issued: January 24, 1978].

108 James Clifford, a talented cartoonist as well as a meticulous technician, in a 30-plus-year career at Waters, personally packed well over 100,000 columns! Charlie [B.S. Chemistry, U. Maine] retired from Lederle Laboratories about 1969 and joined Corning Glass where he developed slurry packing methods and LLPC applications for Controlled Pore Glass [CPG]. He assisted many Corning customers including soon-to-be Waters employees Kenneth Conroe and Gerald Hawk. (See: B.L. Karger, K. Conroe, and H. Engelhardt, *"The Use of Surface Textured Beads for High Speed Column Liquid Chromatography,"* J. Chromatogr. Sci. **8**: 242–250 [1970]; G.L. Hawk, J.A. Cameron, and

B. Larry, *"Chromatography of Biological Materials on Polyethylene Glycol Treated Controlled-Pore Glass"*, *Prep. Biochem. Biotechnol.* **2(2)**: 193–203 [1972]). While living in New Jersey, Charlie was a breeder of iris varieties and served as president of the local chapter of the American Iris Society. When his prized bulbs became infected with a deadly bacterial strain, he reasoned that chlortetracycline [Aureomycin] might cure them. In fact, it did, and he filed a pioneering patent on the control of plant microorganism infections with this antibiotic: C. Pidacks, *"Control of Plant Diseases,"* U.S. Patent 2,720,727 [1955]. In Massachusetts, Charlie's wife Sylvia, formerly a microbiologist at Lederle, who took time off to raise their children, returned to the workforce at Weston Nurseries in Hopkinton. With Charlie's help, she established there their first tissue-culture laboratory to expedite the propagation of new plant species; this activity greatly enhanced the leadership position that Weston Nurseries maintains in New England. When Charlie retired from Waters, he received an unusual parting gift—a Waters Model 244 LC System (like the one shown in the photo in Figure 8–4) so he could continue doing HPLC in his basement! Charlie, now 90, and Sib still live on their Ashland 'farm'.

Page 24

109 K. Unger and H. Gotz, *"Porous Carbon Support Materials Useful in Chromatography and Their Preparation,"* U.S. Patent 4,225,463 [1980].

110 J.H. Knox and M.T. Gilbert, *"Preparation of porous carbon,"* U.S. Patent 4,263,268 [1981].

111 (a) K. Unger and J. Schick-Kalb, *"Preparation of Organically Modified Silicon Dioxides,"* U.S. Patent 4,017,528 [1977]. This patent was filed on 11 November 1974, one year after the corresponding *German Patent* 2357184. (b) K.K.Unger, N. Becker, and P. Roumeliotis, *"Recent Developments in the Evaluation of Chemically Bonded Silica Packings for Liquid Chromatography,"* *J. Chromatogr.* **125**: 115–127 [1976] . By coincidence, this paper immediately precedes a citation classic in reversed-phase LC theory: C. Horváth, W. Melander, and I. Molnár, *"Solvophobic Interactions in Liquid Chromatography with Nonpolar Stationary Phases,"* *Ibid.* 129–156.

112 Sodium silicate [Na_2SiO_3, known as water glass or liquid glass] is made by fusing silicon dioxide [sand] with sodium carbonate. It is highly soluble in water. Acidification of a water solution of sodium silicate produces silicic acid. This, in turn, is precipitated, dried, and subjected to high temperature to create silica gel. Whether created in large cakes [later ground to make small particles] or in sols, silica gel may retain the metal impurities in the original starting materials.

113 M. Kele and G. Guiochon, *"Repeatability and reproducibility of retention data and band profiles on reversed-phase liquid chromatographic columns. II. Results obtained with Symmetry C_{18} columns,"* *J. Chromatogr. A* **830(1)**: 55–79 [1999] [Note: This paper has been cited over 50 times since its publication.]

114 Waters Alliance® HPLC Systems. See: P.D. McDonald, *"James Waters and His Liquid Chromatography People,"* *Waters Whitepaper* [2007]; PDF available: search for **WA62008** on waters.com .

115 Z. Jiang, R.P. Fisk, J. O'Gara, T.H. Walter, K.D. Wyndham, *"Porous Inorganic/Organic Hybrid Particles for Chromatographic Separations and Process for Their Preparation,"* U.S. Patent 6,686,035 B2 [2004]; *Ibid.*, U.S. Patent 7,223,473 [2007].

116 U.D. Neue, T.H. Walter, B.A. Alden, Z. Jiang, R.P. Fisk, J.T. Cook, K.H. Glose, J.L. Carmody, J.M. Grassi, Y-F. Cheng, Z. Lu, and R.J. Crowley, *"Use of high-performance LC packings from pH 1 to pH 12,"* Am. Lab. **32(22)**: 36–39 [1999].

117 By comparison, titration shows that Spherisorb silanols are acidic and Symmetry silanols are neutral. See: A. Méndez, E. Bosch, M. Rosés, and U.D. Neue, *"Comparison of the acidity of residual silanol groups in several liquid chromatography columns,"* *J. Chromatogr. A* **986**: 33–44 [2003].

Page 25

118 J.H. Knox, *"Practical Aspects of LC Theory,"* *J. Chromatogr. Sci.* **15**: 352–364 [1977].

119 (a) Y_F. Cheng, Z. Lu, and U. Neue, *"Ultrafast liquid chromatography/ultraviolet and liquid chromatography/tandem mass spectrometric analysis,"* *Rapid Commun. Mass Spectrom.* **15**: 141–151 [2001]; (b) See reference 177, pp. 201–202; (c) J. Ayrton, G.J. Dear, W.J. Leavens, D.N. Mallet, and R.S. Plumb, *"Use of generic fast gradient liquid chromatography-tandem mass spectroscopy in quantitative bioanalysis,"* *J. Chromatogr. B* **709(2)**: 243–254 [1998]; (d) L. Romanyshyn, P.R. Tiller, and C.E.C.A. Hop, *"Bioanalytical applications of 'fast chromatography' to high-throughput liquid chromatography/tandem mass spectrometric quantitation,"* *Rapid Commun. Mass Spectrom.* **14**: 1662-1668 [2000]; (e) L. Romanyshyn, P.R. Tiller, R. Alvaro, A. Pereira, and C.E.C.A. Hop, *"Ultra-fast gradient vs. fast isocratic chromatography in bioanalytical quantification by liquid chromatography/tandem mass spectrometry,"* *Ibid.* **15**: 313–319 [2001].

120 Reprinted from p. 2 of: *"Intelligent Speed (IS™) Columns,"* *Waters Brochure* 8 pp. [2004]; PDF available, search for **720000577EN** on waters.com .

121 D. T.-T. Nguyen, D. Guillarme, S. Rudaz, and J-L. Veuthey, *"Fast analysis in liquid chromatography using small particle size and high pressure,"* *J. Sep. Sci.* **29**: 1836–1848 [2006].

122 Years earlier, frictional heating concerns were raised by Halász and Guiochon. (a) Halász argued that the upper pressure limit in HPLC should be 500 bar [7400 psi]. See: I. Halász, R. Endele, and J. Asshauer, *"Ultimate Limits in High-Pressure Liquid Chromatography,"* *J. Chromatogr.* **112**: 37–60 [1975]; (b) Guiochon derived a simple procedure for calculating column parameters [dimensions, particle diameter] to gain the necessary efficiency at the optimum flow rate. See: M. Martin, C. Eon, and G. Guiochon, *"Study of the Pertinency of Pressure in Liquid Chromatography. III. A Practical Method for Choosing the Experimental Conditions in Liquid Chromatography,"* *J. Chromatogr.* **110**: 213–232 [1975]; (c) Though a brilliant theoretician, Georges Guiochon always had a very practical focus; *e.g.*, see: G. Guiochon, *"Optimization in Liquid Chromatography,"* Chapter 1 in: C. Horváth, Ed. *High-Performance Liquid Chromatography. Advances and Perspectives* **Vol. 2**, Academic Press, New York [1980].

123 J.E. MacNair, K.C. Lewis, and J.W. Jorgenson, *"Ultrahigh-Pressure Reversed-Phase Liquid Chromatography in Packed Capillary Columns,"* *Anal. Chem.* **69(6)**: 983–989 [1997].

124 (a) K.D. Wyndham, J.E. O'Gara, T.H. Walter, K.H. Glose, N.L. Lawrence, B.A. Alden, G.S. Izzo, C.J. Hudalla, and P.C. Iraneta, *"Characterization and Evaluation of C_{18} HPLC Stationary Phases Based on Ethyl-Bridged Hybrid*

Organic/Inorganic Particles," *Anal. Chem.* **75(24)**: 6781–6788 [2003]; (b) K.D. Wyndham, T.H. Walter, P.C. Iraneta, U.D. Neue, P.D. McDonald, D. Morrison, and M. Baynham, "*A Review of Waters Hybrid Particle Technology. Part 2. Ethylene-Bridged [BEH Technology™] Hybrids and Their Use in Liquid Chromatography*," *Waters Whitepaper* 8 pp. [2004]; PDF available, search for **720001159EN** on waters.com .

Page 26

125 J. S. Mellors and J.W. Jorgenson, "*Use of 1.5-μm Porous Ethyl-Bridged Hybrid Particles as a Stationary-Phase Support for Reversed-Phase Ultrahigh-Pressure Liquid Chromatography*," *Anal. Chem.* **76(18)**: 5441–5450 [2004].

126 I first saw these photos in the award address Jim gave at the Division of Analytical Chemistry symposium in his honor, Fall ACS National Meeting, Boston, August 2007. Jim is the consummate professor for two reasons: (1) you may glean the first from published evidence—he has built a collegial infrastructure within his group for the long haul, training his students in the mechanical arts as well as in analytical chemistry, giving them the necessary tools to build sophisticated, often miniaturized apparatus used in their thesis research, and making sure that those closer to graduation pass on their knowledge to incoming degree candidates [Douglas Wittmer, personal communication to P.D. McDonald]; (2) not to be overlooked among the host of prestigious awards on his resumé are those—perhaps the most satisfying of all—that recognize him for his excellence in and commitment to teaching.

Page 27

127 Extensive information on the current version of this system and column technology may be found respectively at: http://www.waters.com/acquity and http://www.waters.com/acquitycolumns .

Page 28

128 (a) "*ACQUITY UPLC Columns. More Choices. More Information*": PDF available, search for **720001140en** on waters.com; also, check waters.com regularly for the most recent product information and applications; (b) P.D. McDonald, D. McCabe, B.A. Alden, N. Lawrence, D.P. Walsh, P.C. Iraneta, E. Grumbach, F. Xia, and P. Hong, "*Topics in Liquid Chromatography. Part 1. Designing a Reversed-Phase Column for Polar Compound Retention*," *Waters Whitepaper* 8 pp. [2007]; PDF available, search for **720001889EN** on waters.com .

129 J.C. Giddings, "*Comparison of Theoretical Limit of Separating Speed in Gas and Liquid Chromatography*," *Anal. Chem.* **37(1)**: 60–63 [1965]; The Knox hypothesis was first set forth for GC in: J.H. Knox, "*The speed of analysis by gas chromatography*," *J. Chem. Soc.* 433–441 [1961].

130 Experimental conditions: Mobile phase: CH_3CN/H_2O 7/3; Flow rate range: 0.025–3.0 mL/min; Temperature: 30 °C; Solute: Heptanophenone; Instrument: ACQUITY UPLC with TUV [sub-2 μm]; Alliance® [3.5 & 5 μm packings]; Packings: [from 3 manufacturers] 5 brands of sub-2-μm C_{18}-bonded and end-capped porous silica or BEH packings in eleven columns; three brands, one each in three 5-μm columns; three brands, one each in 3.5-μm columns.

Page 29

131 Figures 10–3 through 10–9 were first presented by: U.D. Neue "*The Interrelationship between Pressure and Column Performance*," Abstract 1760-1, *Symposium on High Speed Liquid Chromatography* [chair: P.W. Carr], Pittcon, Chicago [11 March 2009]; Gary Izzo performed the experiments summarized in Figures 10–3, 10–4, 10–5, 10–8, and 10–9.

132 (a) G. Desmet, D. Clicq, and P. Gzil, "*Geometry-Independent Plate Height Representation Methods for the Direct Comparison of the Kinetic Performance of LC Supports with a Different Size or Morphology*," *Anal. Chem.* **77(13)**: 4058–4070 [2005]; (b) G. Desmet, D. Clicq, D.T.-T. Nguyen, D. Guillarme, S. Rudaz, J-L. Veuthey, N. Vervoort, G. Torok, D. Cabooter, and P. Gzil, "*Practical Constraints in the Kinetic Plot Representation of Chromatographic Performance Data: Theory and Application to Experimental Data*," *Anal. Chem.* **78(7)**: 2150–2162 [2006]; (c) See also: http://wwwtw.vub.ac.be/CHIS//research/tmas2/kineticplot/kineticplot.html .

Page 30

133 Reference 27d, pp. 42–49.

134 H. Poppe, "*Some reflections on speed and efficiency of modern chromatographic methods*," *J. Chromatogr.* **778(1–2)**: 3–21 [1997].

Page 31

135 Transcribed from Dr. Anton Jerkovich, Novartis, video interview with Waters, 2004. Dr. Jerkovich did his doctoral studies with Prof. James Jorgenson, U. North Carolina.

136 E. Chambers, D.M. Wagrowski-Diehl, Z. Lu, and J.R. Mazzeo, "*Systematic and comprehensive strategy for reducing matrix effects in LC/MS/MS*," *J. Chromatogr. B* **852**: 22–34 [2007].

137 C.J. Messina, E.S. Grumbach, and D.M. Diehl, "*A Systematic Approach Towards UPLC Method Development*," *Waters Application Note* 5 pp. [2007]; PDF available, search for **720002338en** on waters.com .

138 P.D. Rainville and R.S. Plumb, "*Increasing Sensitivity of Bioanalytical Assays Utilizing Microbore UPLC and Tandem Quadrupole Mass Spectrometry*," *Waters Application Note* 4 pp [2009]; PDF available, search for **720002842en** on waters.com .

139 P. Hong, T.E. Wheat, and D.M. Diehl, "*Analysis of Physiological Amino Acids with the MassTrak™ Amino Acid Analysis Solution*," *Waters Applications Note* 6 pp. [2009]; PDF available: search for **720002903en** on waters.com .

140 L.J. Calton, S.D. Gillingwater, G.W. Hammond, and D.P. Cooper, "*The Analysis of 25-Hydroxyvitamin D in Serum Using UPLC/MS/MS*," *Waters Application Note* 4 pp. [2008]; PDF available, search for **720002748en** on waters.com .

141 (a) R.Lee, M. Roberts, A. Paccou, and M. Wood, "*Development of a New UPLC/ MS Method for Systematic Toxicological Analysis*," *Waters Application Note* 7 pp [2009]; PDF available, search for **720002905en** on waters.com; (b) L.G. Apollonio, D.J. Pianca, I.R. Whittall, W.A. Maher, and J.M. Kyd, "*A demonstration of the use of ultra-performance liquid chromatography–mass spectrometry [UPLC/MS] in the determination of amphetamine-type substances and ketamine for forensic and toxicological analysis*," *J. Chromatogr. B* **836**: 111–115 [2006].

142 J. Shia and D. Diehl, "*Protecting the Food Supply: Rapid, Specific Analysis of Melamine and Cyanuric Acid in Infant Formula by LC/MS/MS*," *Waters Application Note* 9 pp [2008]; PDF available, search for **720002865en** on waters.com .

143 H.B. Hewitson, T.E. Wheat, and D.M. Diehl, "*Application of UPLC Amino Acid Analysis Solution to Foods and Feeds*," *Am. Lab.* **41(2)**: 22, 24, 26 [2009].

144 J. Morphet and P. Hancock, "*A Rapid Method for the Screening and Confirmation of Over 400 Pesticide Residues in Food*," *Waters Application Note* 12 pp [2008]; PDF available, search for **720002628en** on waters.com .

145 (a) M. Petrovic, M. Gros, and D. Barceló, "*Multi-residue analysis of pharmaceuticals in wastewater by ultra-performance liquid chromatography–quadrupole–time-of-flight mass spectrometry*," *J. Chromatogr. A* **1124**: 68–81 [2006]; (b) A.L. Batt, M.S. Kostich, and J.M. Lazorchak, "*Analysis of Ecologically Relevant Pharmaceuticals in Wastewater and Surface Water Using Selective Solid-Phase Extraction and UPLC-MS/MS*," *Anal. Chem.* **80(13)**: 5021–5030 [2008]; (c) B. Kasprzyk-Hordern, R.M. Dinsdale, and A.J. Guwy, "*Multiresidue methods for the analysis of pharmaceuticals, personal care products and illicit drugs in surface water and wastewater by solid-phase extraction and ultra-performance liquid chromatography–electrospray tandem mass spectrometry*," *Anal. Bioanal. Chem.* **391**: 1293–1308 [2008].

146 A. Gledhill, A. Kärrman, I. Ericson, B. van Bavel, G. Linström, and G. Kearney, "*Analysis of Perfluorinated Compounds [PFCs] on the ACQUITY UPLC System & the Quattro Premier™ XE in Es-MS/MS*," *Waters Application Note* 7 pp [2007]; PDF available, search for **720001761en** on waters.com .

147 M. Mezcua, A. Agüera, J.L. Lliberia, M.A. Cortés, B. Bagó, and A.R. Fernández-Alba, "*Application of ultra-performance liquid chromatography–tandem mass spectrometry to the analysis of priority pesticides in groundwater*," *J. Chromatogr. A* **1109(2)**: 222–227 [2006].

148 A.B. Chakraborty, W. Chen, and J.C. Gebler, "*Characterization of Reduced Monoclonal Antibody by On-Line UPLC–UV/ESI–TOF MS*," *Waters Application Note* 4 pp. [2009]; PDF available, search for **720002919en** on waters.com .

149 J.Castro-Perez, K. Yu, J. Shockcor, H. Shion, E. Marsden-Edwards, and J. Goshawk, "*Fast and Sensitivie In Vitro Metabolism Study of Rate and Routes of Clearance for Ritonavir Using UPLC Coupled with the XEVO™ QTof MS System*," *Waters Application Note* 7 pp [2009]; PDF available, search for **720003025en** on waters.com .

150 H. Xie and M. Gilar, "*High Sensitivity Peptide Analysis with the ACQUITY UPLC System*," *Waters Application Note* 4 pp [2008]; PDF available, search for **720002718en** on waters.com .

151 (a) P.A. Guy, I. Tavazzi, S.J. Bruce, Z. Ramadan, and S. Kochhar, "*Global metabolic profiling analysis on human urine by UPLC–TOFMS: Issues and method validation in nutritional metabolomics*," *J. Chromatogr. B* **871(2)**: 253–260 [2008]; (b) I.D. Wilson, J.K. Nicholson, J. Castro-Perez, J.H. Grainger, K.A. Johnson, B.W. Smith, and R.S. Plumb, "*High Resolution 'Ultra Performance' Liquid Chromatography Coupled to oa-TOF Mass Spectrometry as a Tool for Differential Metabolic Pathway Profiling in Functional Genomic Studies*," *J. Proteome Res.* **4(2)**: 591–598 [2005].

152 R.S. Plumb, K.A. Johnson, P. Rainville, B.W. Smith, I.D. Wilson, J.M. Castro-Perez, and J.K. Nicholson, "*UPLC/MSE; a new approach for generating molecular fragment information for biomarker structure elucidation*," *Rapid Commun. Mass Spectrom.* **20(113)**: 1989–1994 [2006].

153 T.E. Wales, K.E. Fadgen, G.C. Gerhardt, and J.R. Engen, "*High-Speed and High-Resolution UPLC Separation at Zero Degrees Celsius*," *Anal. Chem.* **80(17)**: 6815–6820 [2008].

154 This HDMS system is capable of not only measuring the mass, but also revealing the conformation of molecules. Learn more at: http://www.waters.com/synapt .

155 Imatinib, a small molecule tyrosine kinase inhibitor, was introduced in May 2001 by Novartis with the tradenames Gleevec/Glivec. That same month, a TIME magazine cover story hailed Gleevec as a 'magic bullet' cancer cure. This wonder drug exemplifies rational drug design. See: B.J. Druker, "*STI571 [Gleevec] as a paradigm for cancer therapy*," *Trends Mol. Med.* **8(4)**: S14–S18 [2002]. Its creators have received *2009 European Inventor of the Year Awards*; see: http://www.novartis.com/newsroom/media-releases/en/2009/1314511.shtml .

156 R.E. Jacob, T. Pene-Dumitrescu, J. Zhang, N.S. Gray, T.E. Smithgall, and J.R. Engen, "*Conformational disturbance in Abl kinase upon mutation and deregulation*," *PNAS* **106(5)**: 1386–1391 [2009].

Page 32

157 Several references and application notes are available at: http://www.waters.com/patrol . See, *e.g.*: A.S. Rathore, R. Wood, A. Sharma, and S. Dermawan, "*Case study and application of process analytical technology (PAT) towards bioprocessing: II. Use of ultra-performance liquid chromatography (UPLC) for making real-time pooling decisions for process chromatography*," *Biotechnol. Bioengin.* **101(6)**: 1366–1374 [2008].

158 Thomas Alva Edison [1847–1931]: "*If we did all the things we are capable of doing, we would literally astound ourselves.*" Robert Browning voiced a related thought in *Andrea del Sarto, the Faultless Painter* [1855], "*Ah, but a man's reach should exceed his grasp, Or what's a heaven for?*"

Page 33

159 To learn more about the significance of extra-column band spreading, read, *e.g.*, pp. 349–352 in reference 27d.

Page 34

160 Fick [1829–1901] also built the first successful contact lens, testing it first on rabbits, then on himself, and finally on some volunteers. See: http://en.wikipedia.org/wiki/Adolf_Fick .

161 M.Z. El Fallah and M. Martin, "*Influence of the Peak Height Distribution on Separation Performances: Discrimination Factor and Effective Peak Capacity*," *Chromatographia* **24**: 115–122 [1987].

162 For a discussion of the relative merits of resolution and the discrimination factor, see pp. 302–308 in reference 27d.

Page 35

163   W. Trappe, "*Separation of Biological Fats from Natural Mixtures by Means of Adsorption Columns. I. The Eluotropic Series of Solvents,*" *Biochem. Z.* **305**: 150–161 [1940].

164   See pp. 192-197 in reference 80.

165   R. Neher, pp. 75-86 in: G.B. Marini-Bettòlo, Ed., *Thin-Layer Chromatography [Proceedings of the Symposium held at the Istituto Superiore di Sanita, Rome, 2-3 May 1963]*, Elsevier, Amsterdam [1964].

Page 37

166   See reference 27d, pp. 14–15 for a discussion of this topic.

Page 38

167   The five-sigma method was first proposed by Dr. Richard King, a senior scientist in Waters R&D, in the late 1970s.

Page 39

168   For a more detailed discussion, with diagrams and early references, of both recycle techniques, see reference 28, pp. 90–91.

169   P.D. McDonald and H.S. Schultz, "*Ultra High Efficiency Size Separations of Small Organic Molecules,*" *Poster* **III–6**, HPLC 1982, Cherry Hill, New Jersey. The late Herman Schultz had given me a pair of the first prototype 500-Å Ultrastyragel™ columns, each with 25,000 plates/30 cm, for these experiments. I presented this poster on June 10th, 1982, the same day my wife Kathryn and I celebrated our 15th wedding anniversary. Prof. Jim Jorgenson had delivered the first lecture that morning, entitled "*Liquid Chromatography in Open-Tubular Columns.*" Kathy and I celebrated our 35th anniversary during HPLC 2002, coincidentally held in Montréal, the location of our 1967 honeymoon!

170   Jim reported this work at the Fall ACS Meeting, Boston, 2007. See this initial publication: K. Lan and J. Jorgenson, "*Pressure-Induced Retention Variations in Reversed-Phase Alternate-Pumping Recycle Chromatography,*" *Anal. Chem.* **70(14)**: 2773–2782 [1998].

171   For some novel ideas on the mechanism of reversed-phase separations, see: P.D. McDonald, "*Improving Our Understanding of Reversed-Phase Separations for the 21st Century,*" Chap. 7 in: *Adv. Chromatogr.* **42**: 323–375 [2003].

Page 40

172   For more information, visit: http://www.waters.com/uplc . See also, a list of resources for UPLC technology in Chapter 14.

Page 41

173   "*Elution Chromatography,*" Chapter 15, and "*Paper Chromatography of Amino Acids,*", Chapter 24 in: L.F. Fieser, *Experiments in Organic Chemistry*, third ed., Heath, New York [1957]; these two experiments were new to this edition. The first and second editions were published in 1935 and 1941, respectively. My copy still bears the charcoal and chemical stains from lab spills in the fall of 1961, my freshman year at Brown U.!

174   L.F. Fieser and M. Fieser, *Organic Chemistry*, third ed., Reinhold, New York [1956]. The first edition was published the year I was born, 1944, and the second edition six years later. In their Preface, the Fiesers stated: "*The idea of including brief biographical sketches of past and present chemists associated with the developments cited was introduced by Dr. Hans Hensel in his German translation of our second edition. The list of biographies is now increased to 454 entries, each of which gives the person's full name and vital statistics, the university at which he earned a higher degree or studied, the professor under whom he worked, and the university or company of his major service. Nobel Prize awards are indicated, and references are given to memorial lectures and biographical sketches.*"

175   W.G. Africa, "*The Glass Sand Industry of the Juniata Valley,*" pp. 36–43 in: *Annual Report of the Secretary of Internal Affairs of the Commonwealth of Pennsylvania. Part III. Industrial Statistics 1885*, **Vol. XIII**, F.K. Meyers, State Printer, Harrisburg [1886]; full document digitized by Google.

176   Though U.S. Silica sold most of its previously acquired assets of the former Floridin company, it still manufactures FLORISIL®, a coprecipitate of magnesia and silica used as a chromatographic adsorbent, first made by the Floridin Company in 1941. John Wannamaker, once manager of sales for Floridin in their Pittsburgh headquarters told me in 1982 that a sizeable percentage of their Florisil business disappeared, never to return, following a visit to his office by U.S. Government agents investigating the regular export of large quantities of Florisil bulk adsorbent from a New York office front to Marseille, France. Apparently, unknown to Floridin, tons of Florisil were being used in black market factories for the purification of morphine, from opium poppy extracts, and/or its derivative heroin [diacetylmorphine].

177   For Uwe's personal career retrospective, please see: U.D. Neue, "*Hollow Sticks with Mud Inside: The Technology of HPLC Columns,*" Chap. 12 in: H. Issaq, Ed., *A Century of Separation Science*, Marcel Dekker, New York [2002].

Vitae

Dr. Patrick Doyle McDonald is a Corporate Fellow in Chemistry Operations, Waters Corporation, where, since 1974, he has done pioneering research and product development in areas as diverse as preparative LC, sample preparation, industry-standard HPLC solvent purity specifications, post-column reaction detection, environmental analysis, mobile-phase degassing technology, and electronic information systems.

Born and raised in Rockville Centre, Long Island, New York, he received his Sc.B. in Chemistry in 1965, with a second major in English literature, from Brown University. In 1970, he was granted a Ph.D. in bioorganic chemistry from the Pennsylvania State University. His thesis, *On the Mechanism of Phenolic Oxidative Coupling Reactions*, was supervised by Prof. Gordon A. Hamilton. From 1970 to 1974, he was an NIH Postdoctoral Research Trainee in Endocrinology, studying steroid biochemistry with Prof. Seymour Lieberman in the Departments of Biochemistry and Obstetrics/Gynecology and the International Institute for the Study of Human Reproduction, College of Physicians and Surgeons, Columbia University.

While at Columbia, in 1972, he bought an ALC 100 system from James Waters and did seminal work in steroid separations using HPLC. His experience with LC techniques, however, especially preparative TLC, open column liquid-liquid partition, and liquid-solid adsorption chromatography, began in 1961. He joined Waters Associates on July 1, 1974, as a senior research chemist, rising to his current role by way of positions such as Director of Chemical R&D and Director of Strategic Development.

He is the principal inventor of radial compression technology, the first really new method of packing LC columns in the then 75-year history of chromatography. He guided the development of the PrepLC™/System 500, the first time that a commercial LC instrument was designed to incorporate a specific, proprietary column technology [Prep-Pak™ Cartridges]. After the second system was placed at Harvard, Prof. Elias J. Corey, in comments published in the *New York Times*, called it *"a revolution in liquid chromatography"* and a *"good example of the fine technological work in American industry."*

Dr. McDonald's work in sample preparation catalyzed a worldwide revolution in analytical methods development. He lead the team which invented in October 1977, rapidly developed, and brought to market for sale in January 1978, the Sep-Pak® Cartridge – the first commercial miniature silica-based column liquid-solid extraction device. In 1996, he and his team invented and brought to market Oasis® Sample Extraction products, containing a patented polymeric sorbent with revolutionary properties for reversed-phase SPE.

Dr. McDonald is the author or co-author of many refereed papers, book chapters, a founding editor and architect of Waters Applications Library offline and online databases, and a frequent contributor at international symposia in the last 40 years. He is a career-long member of the ACS, AAAS, Sigma Xi, and the Chromatographic Society (London). He holds nine U.S. and many foreign patents and has received two Millipore awards for Technical Innovation. In September 2008, as part of the celebration of Waters 50-year legacy, Dr. McDonald was honored for his career achievements by election to the inaugural group of Waters Corporate Fellows, collectively cited for their *"keen understanding, purposeful innovation, and meaningful impact."* He is an accomplished singer and photographer. He resides in a unique hexagonal home of his own design in Holliston, Massachusetts, with his wife Kathryn. Their two sons, Patrick and Brian, are both graduates of Brown University [computer engineering/economics and computer science, respectively].

Dr. Uwe Dieter Neue is a Corporate Fellow and Director of External Research in Chemistry Operations, Waters Corporation, where, since 1976, he has worked on stationary phases, column packing technology, and applications, from sample preparation to method development. Dr. Neue is an expert with more than 30 years of experience in the field of HPLC. He started his research in HPLC during the pioneering period of the early 1970s at the Institute of Applied Physical Chemistry of the University of Saarbrücken. For his thesis, *Chemische Nachweisverfahren in der Hochdruckflüssigkeitschromatographie [Chemical Derivatization Methods in HPLC]*, supervised by his mentors, Prof. István Halász and Prof. Heinz Engelhardt, he was awarded the Doctor of Science degree in 1976.

Dr. Neue's work has been described in nearly 150 papers and book chapters in the field of chromatography, most of them published in the last decade. He has been a frequent contributer and invited speaker at international meetings throughout his career. His definitive book on *HPLC Columns* was published in 1997.[27d] His current research interests include ultra-fast separations, preparative chromatography, sample preparation, stationary phases, and the separation of macromolecules. He is a member of the editorial boards of the *Journal of Chromatography* and the *Journal of Separation Science*. In September 2005, his *"outstanding achievements"* were celebrated by the Hungarian Society for Separation Sciences at the Balaton Symposium, Siófok, where they presented to him the Halász Medal, an award established in 1997 to commemorate Dr. Neue's late mentor. In September 2008, Dr. Neue was also honored for his career achievements by election to the inaugural group of Waters Corporate Fellows, collectively cited for their *"keen understanding, purposeful innovation, and meaningful impact."* He resides in Ashland, Massachusetts, with his wife Lynda and their two sons, Alexander and Martin.

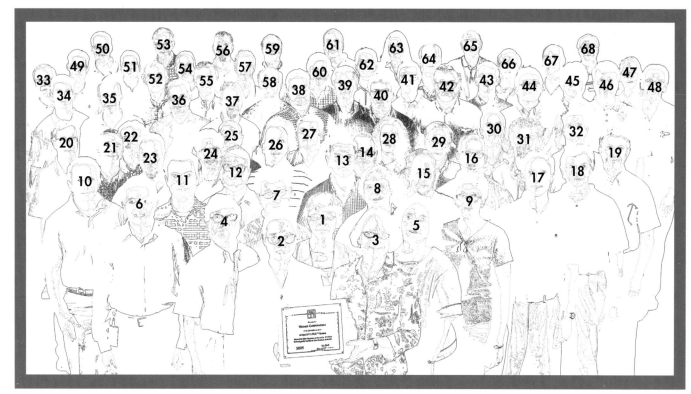

Key to Photograph on outside rear cover

R&D 100 Award-winning principal ACQUITY inventors and developers: 1–Dr. Marianna Kele; 2–Dr. Yuehong Xu; 3–Pamela Iraneta; 4–Rahim Hamid; 5–Karen Thorson; 6–Amit Arora; 7–Maruth Sok; 8–Chris Grzonka; 9–Pamela Longenback; 10–Carlos Gomez; 11–John Lamoureux; 12–Ryszard Feldman; 13–Joseph Kareh; 14–Peter Kirby; 15–Russell Keene; 16–Barry Sunray; 17–Brian Edwardsen; 18–Peyton Beals; 19–Richard Kent; 20–Stanley Pensak; 21–Edwin Denecke; 22–Charlie Murphy; 23–Marc Lemelin; 24–Dr. Uwe Neue; 25–David Prentice; 26–Jon Belanger; 27–John Auclair; 28–Dr. Aisling O'Connor-Stevenson; 29–Dennis DellaRovere; 30–Eric Grumbach; 31–Douglas Wittmer; 32–Andrew Michaud; 33–Joseph DeLuca; 34–Aaron LeBeau; 35–Steven Binney; 36–Joseph Luongo; 37–Raymond Fisk; 38–James Usowicz; 39–John Angelosanto; 40–Miguel Soares; 41–Dr. Thomas Walter; 42–Kenneth Plant; 43–Dr. Kevin Wyndham; 44–Robert Dumas; 45–David Piper; 46–Mark Robinson; 47–Daniel McCormick; 48–Dr. Joseph Jarrell; 49–Ed Ognibene; 50–Dr. John O'Gara; 51–Bruce Smith; 52–Theodore Dourdeville; 53–Dr. Anthony Gilby; 54–Dr. Jeffrey Mazzeo 55–Ted Ciolkosz; 56–Dr. Richard Andrews; 57–John Heden; 58–Joseph Antocci; 59–Jose DeCorral; 60–John Leason; 61–Mike LeBeau; 62–Frank Rubino; 63–Jeanine Pippitt; 64–John Maillett; 65–Michael Savaria; 66–Steven Ciavarini; 67–Paul Linderson; 68–Frank Denecke;

Missing: Jennifer Burzynski; David Friswell; Scott McLaren; Mark Moeller